THE ELEMENTS OF
TECHNICAL WRITING

THE
ELEMENTS
OF TECHNICAL
WRITING

Gary Blake and Robert W. Bly

New York   San Francisco   Boston
London   Toronto   Sydney   Tokyo   Singapore   Madrid
Mexico City   Munich   Paris   Cape Town   Hong Kong   Montreal

Longman Publishers
1185 6th Avenue
New York, NY 10036

Library of Congress Cataloging-in-Publication Data

Bly, Robert W.
The elements of technical writing / Robert W. Bly and Gary Blake.
p.   cm.
Includes index.
ISBN 0-02-013085-6 (pbk)
1. Technical writing—Handbooks, manuals, etc.   I. Blake, Gary.
II. Title.
T11.B628   1993   92-44998   CIP
808'.0666—dc20

10 9 8 7 6 5 4

Designed by Nancy Sugihara

Printed in the United States of America

03012 8555

To my parents
R.B.

To Bonnie Blake-Drucker
G.B.

Contents

vii

3 A Few Useful Rules of Punctuation, Grammar, Abbreviation, and Capitalization   47

5 Words and Phrases Commonly Misused in Technical Writing 75

PART II
TASKS OF THE TECHNICAL WRITER 97

6 Proposals and Specifications 99

Acknowledgments

Although we are acknowledged as the authors, we must thank those who have lent help, inspiration, and encouragement to this book.

Thanks to Amy Bly, Eve Blake, Tim Yohn, and the late Jack Nathan. Thanks also to Richard Nelson for his help with aspects of this book pertaining to the world of systems.

We are indebted to our agent, Dominick Abel, and to our editor, Natalie Chapman. We also recognize the inspiration of Strunk and White as well as that of our clients. By working with our clients—through seminars, writing projects, and tutorials—we were able to find the examples, illustrations, and other materials that helped ground our discussions in the work of a variety of engineers and technicians.

Introduction

In 1981, we wrote *Technical Writing: Structure, Standards, and Style*. It was our first book; it was sold by our newly acquired agent; it was accepted by the first publisher who read it.

All of us thought the book would sell modestly. It surprised us by going through ten printings in ten years.

After the book was published, we started our respective businesses and began to get a lot more experience working with the memos, reports, proposals, and manuals produced by technical writers. We also watched certain technologies expand exponentially. Now, finally, we have the opportunity to present what we've learned, expand our horizons, and write what we hope is a book that meets the changing technical-writing needs of the 1990s.

So, here we are with a new publisher, the same agent, and a book that benefits from ten years of collected experience, wisdom, and commentary from readers in a variety of fields, including manufacturing, aerospace, telecommunications, electronics, food service, waste management, banking, insurance, chemical processing, computers, software, law, health care, and cable television.

We believe you'll find that *The Elements of Technical Writing* has some important features that point up the vast business and technological changes that have taken place in the past ten years.

To begin with, we've included a special section with examples and text that address the specialized writing problems of systems professionals. We now feel able to define and lay down principles about a type of writing that changed radically throughout the 1980s. We have tackled specific problems of punctuation in the systems environment and have tried to define the types of manuals written by systems analysts and software engineers.

By reformatting our book, we have been able to expand our commentary on a variety of documents written by technical professionals, especially proposals, reports, manuals, letters, and memos. Thanks to our clients, we're able to offer readers many "real-life" examples drawn from these documents.

Last, we have honed our definitions, tips, examples, and sentences to make them sharper, clearer, and more authoritative. We have refined our thinking and placed new emphasis on the technologies of the 1990s.

Here's how the book is organized:

Chapter 1 provides a checklist of the 10 qualities that an effective technical document should have. Some of these also apply to nontechnical writing, but examples and text explain the special relationship of these qualities to the technical composition.

Chapter 2 covers the proper use of numbers, units, equations, and symbols. It will help keep mathematics correct and consistent in your writing.

Chapter 3 covers the basics of grammar—with emphasis on rules that help you avoid the most common errors in technical writing.

Chapter 4 presents the basic principles of technical communication. It discusses the tone, style, and voice used in the various types of technical writing.

Chapter 5 shows you how to avoid jargon, clichés, antiquated phrases, and other evils.

Chapter 6 discusses proposals, with emphasis on viewing them as "selling documents."

Chapter 7 discusses the structure and style of technical reports.

Chapter 8 covers letters and memos, with information on format, wording, salutations, and closings.

Chapter 9, about manuals, focuses on documentation while

acknowledging that manuals are a staple of many technical fields.

The appendix covers an area of vital concern for many technical writers: technical writing for information systems professionals.

The 1990s will be an era of continued growth in the sciences, as recent advances in genetics, data communications, computer technology, medicine, energy sources, and ecology are superseded by new advances. The next 20 years promise even greater growth, hundreds of thousands of new jobs in the sciences, and an increased emphasis on clear, concise, accurate scientific writing.

We hope that you'll find *The Elements of Technical Writing* enjoyable as well as handy. We tried to follow our own precepts, keeping the book brief, lively, and to the point.

Many other thoughts are in our minds as we write this introduction: the perils of writing about writing, our hope that we have covered your own technical field in the examples, our concern that we have steered a reasonable course between being authoritative and being open-minded. But, for the sake of brevity, we'll stop here.

PART I

Elements
of Technical
Writing

1

Fundamentals of Effective Technical Writing

> *"Newspaper reporters and technical writers are trained to reveal almost nothing about themselves in their writing. This makes them freaks in the world of writers, since almost all of the other ink-stained wretches in that world reveal a lot about themselves to the reader."*
>
> —Kurt Vonnegut, Jr., novelist and former technical publicist for General Electric

What Is Technical Writing?

Most technical writers would hardly classify themselves as "freaks in the world of writers." Yet technical writing—the literature of science, technology, and systems development—is different from journalism, fiction, advertising copywriting, essays, plays, poetry, and other nontechnical prose.

Just what is technical writing? Technical writing is defined by its subject matter: It is writing that deals with topics of a technical nature. By *technical* we mean anything having to do with the specialized areas of science and technology.

Traditionally, technical writers were thought of as "engineering writers." And technical writers continue to be employed in such engineering-oriented industries as aerospace, defense, consumer electronics, chemical processing, pulp and paper, mining, construction, fiber optics, instrumentation and controls, and many other fields.

With the tremendous increase in the use of computers, many

3

of the jobs for technical writers today are in computers and related high-tech areas, where writers are needed to produce software documentation, user's manuals, and a variety of other technical documents.

In addition to engineering and information systems, technical writers are involved in all areas of physical, natural, and social sciences, including anthropology, archaeology, biology, biochemistry, biotechnology, botany, earth science, ecology, genetics, geology, linguistics, management science, medicine, microbiology, psychiatry, psychology, sociology, statistics, virology, and zoology, to name a few.

Technical vs. nontechnical writing

Because technical writing usually deals with an object, process, system, or abstract idea, the language is utilitarian, stressing accuracy rather than style. The tone is objective; the technical content, not the writing style or author's feelings toward the subject, is the focal point.

The difference between technical writing and ordinary composition is more than just content, however. The two differ in purpose as well. The primary goal of any technical communication is to *transmit technical information accurately.* In this regard, technical writing differs from popular nonfiction, in which the writing is meant to entertain, or from advertising copywriting, which is intended to sell.

Technical writers are concerned with communication. If they have to, they will sacrifice style, grace, and technique for clarity, precision, and organization.

Good Technical Writing Is . . .

1. Technically Accurate

Because the purpose of technical writing is the transmission of technical information, even the most beautifully written technical document is a failure if the facts, theories, data, and observations presented are in error. The content must be true and as technically accurate as possible.

At best, technical writing that contains inaccurate statements communicates to your readers that you were too lazy or uncaring to research your topic or check your facts. At worst, it destroys your credibility. Readers spot the errors and conclude you don't know what you are talking about.

Another reason why accuracy is so important in technical writing is that technical documents, unlike articles or books read for pleasure, are not merely leisure reading: Your readers make business decisions, operate equipment, or draw scientific conclusions based on the information you present. So accuracy is essential.

For example, if an article in the newspaper accidentally describes the distance from the earth to the sun as 9.2 million miles instead of 92 million miles, no real harm is done. Readers may be embarrassed when they use the erroneous figure in casual conversation, but that's about the worst that can happen. They will not blow up a rocket or program the wrong flight plan into NASA's computer; the information has been presented for entertainment and enlightenment only.

On the other hand, let's say an operating manual for a pump inaccurately lists the maximum operating pressure as 500 psig instead of 50 psig. The operator pushing the pump past its limit could be hurt or even killed, and you could lose thousands of dollars if the accident shuts down your processing plant.

Errors in technical documents can cost industry millions of dollars, and the results of good scientific work can be obscured by hastily prepared reports that are full of inaccuracies.

Documents prepared by technical writers or editors should always be reviewed for accuracy by one or more scientists or engineers familiar with the technology. Calculations and tables of numbers should be checked several times; errors in numbers are far less glaring to the nontechnical proofreader than errors in spelling or grammar.

Documents prepared by technical experts should be reviewed

by a colleague for accuracy as well as by a technical editor who checks for transposed figures, typos, and things of that nature.

2. Useful

In most cases, people read a technical report, data sheet, manual, or proposal because they intend to use the information in some way—either to make a purchase decision, further their own research, or operate a piece of equipment.

Make sure every sentence contains useful information. Delete from your manuscript writing that is ornamental, that entertains but does not inform, or that presents interesting but unnecessary background information. While you may be in love with a particular passage or find a certain fact fascinating, your readers probably don't have time for it. They want you to tell them just what they need to know about the topic—no more, no less.

One example of the wrong approach is a brochure produced by a manufacturer of industrial gases. Aimed at process engineers, it was designed to communicate the many innovative services the manufacturer offered to help ensure a consistent, reliable supply of high-quality gas to meet the reader's process requirements. Unfortunately, this useful information was buried in the second half of the brochure; the introduction was a long-winded history of the manufacturer and its various divisions, acquisitions, and changes in corporate structure—something of interest to the manufacturer but not to its customers. The best strategy in this case would be to delete or substantially reduce the unnecessary corporate "puff," relegate it to the back cover, and put the "meat" of the story up front.

In the same way, a user's manual shouldn't tell how the software was invented, how many lines of code it contains, or what language it is written in—unless this information is important to the user. It should simply and directly tell your users that to activate function X, they have to press button Y.

3. Concise

If your readers are like most people in industry, they are extremely busy. Concise technical writing is easier and less time-consuming to read than wordy technical writing.

Also, long-winded technical writing costs money: If by being concise you can reduce your user's manual from 200 to 50 pages, you'll save on printing costs.

The main reason most technical documents are too long is that deleting the wordiness and repetition is hard work—and working at their writing is something most technical professionals want to avoid. As Blaise Pascal once wrote, "I have made this a long letter because I haven't the time to make it shorter."

By avoiding undesirable repetition, wordy phrases, rambling explanations, pompous language, and jargon, you can produce succinct, readable prose without sacrificing technical content. Your words and phrases should be precise, your composition well organized and to the point. The text in the left column, written by an engineer, is wordy and full of needless jargon; a more concise version is shown at right.

It is also essential that the interior wall surface of the conduit be maintained in a wet condition, and that means be provided for wetting continually the peripheral interior wall surface during operation of the device, in order to avoid accumulation of particulate matter about the peripheral interior surface area.	The interior wall must be wetted regularly so solids do not stick to it.

Here's another example. At left is the lead paragraph from a book on gene control. It consists of a single 79-word sentence. At right the same thought is expressed in a sentence of only 16 words.

In this book I have attempted an accurate but at the same time readable account of recent work on the subject of how gene controls operate, a large subject which is rapidly acquiring a central position in the biology of today and which will inevitably become even more prominent in the future, in the efforts of scientists of numerous different specialisms to explain how a single organism can contain cells of many different kinds developed from a common origin.*

This book is about how gene controls operate—a subject of growing importance in modern biology.

Many technical professionals, businesspeople, and scientists have been trained to think that long, jargon-filled reports are more substantial and important than simple, concise writing. They are wrong. Many profound thoughts and concepts can be expressed in brief form: The First Amendment to the United States Constitution contains 45 words, Newton's first law of motion contains 29, and $E = mc^2$ has only five characters.

4. Complete

Many people confuse conciseness with brevity, but the two are different. Being concise means telling the whole story using the fewest possible words. Being brief means keeping it short, regardless of whether this is achieved through careful writing or by deleting sections of text at random.

When writers confuse conciseness with brevity, they often end up cheating the reader: They keep it short by leaving out important stuff. Especially for a writer dealing with a complex subject in which he or she is not well versed, there is a strong

*J.A.V. Butler, *Gene Control in the Living Cell* (New York: Basic Books).

temptation to handle particularly difficult material simply by skipping over it.

Unfortunately, this solution is not only lazy; it's inadequate. Good technical writing is complete, and that means you include everything the reader needs to know, with nothing essential left out.

In particular, be careful when editing specifications, lists of features, or visuals such as charts and graphs. Professional editors and writers tend to lift tables, charts, graphs, and specifications from source material and use them "as is" without further scrutiny. But such material should be handled as carefully as narrative text.

Make sure the specification is complete and that no key data arc omitted. Ask technical experts whether there is another graph or drawing your reader would want to see that isn't there. The document should be clear and concise enough to be of interest to the least technical reader, yet complete enough so that the technically sophisticated readers will find everything they need.

Being complete, however, does not mean you include everything that can be said about the technology. Virtually every technical document in existence today could have been ten times longer; you must be selective as well as thorough. A complete document tells your readers all they need to know about the topic, but not a word more.

5. Clear

In fiction, essays, and other nontechnical writing, clarity is not necessarily the key to good style. Obscurity has its place in some forms of literature, but not in technical writing, where comprehension is the chief concern.

Technical writers succeed when their work is readily understood by the intended audience. Here are a few suggestions for making your writing clearer:

- *Keep the writing short and simple.* Use small words, not big ones. Keep paragraphs and sentences short. Whenever you can logically break a long paragraph into two shorter ones, do so.

Break the writing into short sections, each with a descriptive subhead, and limit each section to one topic, theme, or idea. Sentences, paragraphs, and chapters that express one idea are easiest to understand; one of the most common writing mistakes is to cram too many thoughts into a single sentence or paragraph.

- *Avoid jargon.* One reason why readers may be unable to understand a particular piece of writing is that they do not understand one or more of the words being used. While not understanding terms is not the only cause of poor communication, it's certainly one of the major causes. Some technical terminology is valid and necessary. But too much technical jargon makes writing incomprehensible. Remember, not everyone knows all the latest buzz words in your discipline or industry. And your readers are busy people, more likely to put your report aside than to reach for a technical dictionary.

- *Present your story in a logical, orderly fashion, one step at a time.* Technical professionals and other nonwriters often feel they have to tell every last detail in the sentence they're writing at the moment. The result is too much information in too short a space.

  Although brevity and conciseness are desirable, it's better to spread your information out in a series of well-ordered paragraphs than to lump everything together in a jumble. Too much description in a single sentence makes writing dense and difficult to read; let the reader breathe a little by presenting points as bite-size tidbits. If a particular piece of information is important but doesn't fit logically into the flow of your narrative, place it in a sidebar or appendix.

- *Use visuals.* Whether one picture is actually worth a thousand words is debatable. But it is true that certain information is communicated more effectively in visual form than as narrative text. For example, no written text can ever describe what a zebra looks like as precisely as a photograph. And a blueprint can tell a civil engineer more about the structure of a bridge than sentences and paragraphs can.

Table 1-1. Common types of visuals and what they communicate.

Visual	What It Shows
photograph or drawing	what something looks like
map	where it is located
diagram	how it is put together
flow chart	how it works
graph	how one variable changes in relation to another
pie chart	proportions and percentages
bar chart	comparisons among quantities
table	a body of data
structured diagram	the components of a system and how they interrelate

The different types of visuals and the messages they communicate are listed in Table 1-1.

6. Consistent

"A foolish consistency," wrote Ralph Waldo Emerson, "is the hobgoblin of little minds." Maybe so. But inconsistencies in technical writing confuse readers and convince them that your scientific work is as sloppy and unprofessional as your prose.

Today technical literature is plagued with random and unnecessary capitalization, mixed sets of units of measure, and indiscriminate use of abbreviations, punctuation, and rules of grammar. Consider this example:

These mist eliminators are available in diameters as large as 7 to eight feet in diameter, but are usually limited to 36–48″ diameter in sugar refinery applications. The mesh pads are usually 12 to sixteen cm thick. In most sugar refinery installations, they will eliminate B.O.D. and cut production loss up to ninety-six percent. As you know, BOD is caused by decomposing organic matter in wastewater streams.

Why, asks the puzzled reader, are diameters given in English units (feet and inches) while thickness is given in metric units (centimeters)? Why write B.O.D. in one sentence and BOD in the next? Why write out eight and sixteen when 7 and 12 are in numerals?

The trouble with being inconsistent is that you are automatically wrong at least some of the time.

It takes a careful editor to keep everything in one consistent style. The reward of uniformity is a technical document that reads as if it was written by an educated, literate person concerned with accuracy and clear communication.

7. Correct in Spelling, Punctuation, and Grammar

All writing—with the possible exception of dialect, poetry, and experimental fiction—must follow the rules of spelling, punctuation, and grammar laid down in the *U.S. Government Printing Office Style Manual*, the *New York Times Manual of Style and Usage*, Strunk and White's *Elements of Style*, and other standard references on the English language.

Many scientists, engineers, and systems professionals are not overly concerned with these seemingly picayune matters of English usage. After all, they reason, technical people are interested in science and technology—not periods, parentheses, and participles.

Unfortunately, even the most indifferent readers are quick to spot misspelled words, poor grammar, and incorrect nomenclature in the writing of others. Such errors indicate that authors are lazy or uncaring about their work.

In addition to improving writing skills, *The Elements of Technical Writing* is designed to be used as a style guide, especially by technical writers and technical professionals employed by companies that do not have a corporate or departmental style book to guide them in matters of punctuation, grammar, and usage. If your office has a style guide that's widely used but conflicts with *The Elements of Technical Writing*, you should probably follow your employer's style guide unless you get permission from your supervisor to do otherwise.

The exception would be where the corporate style guide is blatantly in error. The best strategy in such a case is to point out the error and get it corrected so proper usage will be allowed.

8. Targeted

One of the most difficult challenges technical writers face is writing to the level of technical proficiency and understanding of the intended audience.

This would be easy if all readers were at the same level of education and experience, as is sometimes the case. More often, though, you are writing for an audience with diverse backgrounds and levels of understanding of the topic. The question therefore arises: "Should I write primarily for the benefit of the most technically sophisticated readers? Or should I write at a level at which the least educated reader can not only understand but actually enjoy reading the material?"

The concern is that if the writing is too simple, technical experts will feel you are talking down to them and be insulted or turned off. At the same time, there's the risk of being incomprehensible to less educated readers if you write at too high a level of technical complexity.

While this problem does not have a solution that lends itself to rules or formulas—indeed, it comes down to individual judgment rendered on a case-by-case basis—here are some guidelines we have found helpful in tailoring our writing to the intended audience:

- *Define your audience.* Many people who are not professional writers but do some writing as part of their job never stop to think about who their audience is. But how can you communicate clearly if you don't know whom you are talking to? Are you writing to chemical engineers? biochemists? lab technicians? chemistry teachers? chemical engineering students? The same topic slanted toward different audiences may require totally different treatments. Things to look for when analyzing your audience include industry, job title, job function, education, and level of interest in the subject.

• *Picture your audience.* Going one step beyond merely identifying your audience is really putting yourself in their shoes and focusing your writing on their needs, interests, and level of technical understanding. In his book *Writing Nonfiction That Sells*, Samm Sinclair Baker gives this tip to help writers visualize their readers:

"When I write an article or book, I usually clip from a magazine a few photos of individuals whom I consider to be 'typical' readers on the subject. I like to 'see' the people I'm trying to reach as actual visual images propped prominently in sight on my desk. In this way, I find myself better able to 'talk' to them and even 'listen' to them via my writing."

• *Write for the majority while accommodating significant minorities.* A question frequently asked at our writing seminars for engineers is, "If I'm writing for a mixed audience of technical and nontechnical people, should I write for the engineers only, or try to please the nontechnical readers as well?"

The answer is you should write to please the majority while accommodating other readers if they are significant in number or status. Let's say you're writing a technical data sheet. It will be used primarily by engineers to specify equipment, but end users—some of whom may be nontechnical businesspeople or municipal officials—will also look at it when evaluating whether to buy your product. In this case, you should make the data sheet technical enough to give the engineers the information they need to specify your system, but you should also include text descriptions that stress the advantages of your technology in terms the lay reader can understand.

• *Use the "gist" test.* When writing to a mixed audience of readers of varying levels of technical depth, how clear and simple must you be? The "gist" test is a good guide.

This is the gist test: Even though the writing may contain technical terms that only the technically oriented reader would know, nontechnical readers should be able to get at least the *gist* of the story by reading your copy, even if they do not totally understand the finer points of the discussion. Conversely, if nontechnical people reading your copy throw their hands up and say they don't understand a word, chances are you have not expressed yourself clearly and your technical audience will also not understand much of what you've written.

• *When the primary audience is nontechnical, flesh out explanations.* When the primary audience consists mainly of business types, the general public, or other lay readers, it's better to overexplain than to explain too little. Being a patient teacher in your writing ensures that even the least technical of your readers understands you, and being simple does minimal harm to "techies" (few people ever complain that a piece of writing is too easy to read).

The best way to explain things clearly is to take your time. Don't feel you have to explain every technical concept or term in the sentence in which it is first used. Instead, use a new sentence or paragraph to explain the new idea carefully and thoroughly in plain, simple, nontechnical English, as in the example below:

The W-100 Controller is designed to operate in a *compound-loop control mode.*

In the compound-loop control mode, the injection rate of gas into the fluid is controlled by two variables: the concentration of gas in the process stream, measured in milligrams per liter, and the flow rate of water.

The concentration is measured by a residual analyzer, and the flow rate is measured by a flowmeter.

Based on measurement of the concentration and flow rate, the controller changes the position of the valve's actuator to achieve the feed rate necessary to adjust the gas concentration to the desired setpoint.

If ideas require explanations that are cumbersome or overly lengthy, these explanations can be presented in sidebars or appendixes.

- *When the primary audience is technical, make explanations parenthetical.* For the technical audience, you assume a greater depth of understanding and can therefore omit long, detailed explanations of basic concepts the reader already knows.

Keep in mind, however, that it's often dangerous to assume *too* much knowledge. If the reader doesn't understand a concept you've treated as common knowledge, he or she will be lost—and probably will be too embarrassed to ask anyone for clarification. And with the amount of technical knowledge in the world growing at a frantic rate, it's more difficult than ever for technical people to keep up with their field. So don't assume a greater knowledge than the reader has: Many know less than you think they do or they admit they do.

The solution is to keep the discussion technical and concise, yet "sneak in" short explanations that clarify things for the reader who may not fully understand what you're talking about. This is done by treating the explanation almost as an aside: You don't write an essay; instead you present a brief description or short definition that enables the reader to keep up and follow you. This is often separated from the rest of the sentence by parentheses, a comma, or a dash:

> The W-100 Controller is designed to operate in a compound-loop control mode (where the controller adjusts the valve based on the fluid flow rate and the gas concentration).

> The W-100 Controller is designed to operate in a compound-loop control mode, where the controller adjusts the valve based on the fluid flow rate and the gas concentration.

> The W-100 Controller is designed to operate in a compound-loop control mode—a mode of operation where control of the valve is based on measurement of flow rate and gas concentration.

In these examples, compound-loop control is given a one-line definition rather than the lengthier step-by-step explanation used in the previous example.

9. Well Organized

Poor organization is one of the major obstacles preventing people from presenting technical material in a clear and easy-to-follow manner. As Jerry Bacchetti, an experienced technical editor, points out: "If the reader believes the content has some importance to him, he can plow through a report even if it is dull or has lengthy sentences and big words. But if it's poorly organized—forget it. There's no way to make sense of what is written."

Poor organization stems from poor planning. While a contractor would never think of putting up a building without first having an architect or engineer draw a blueprint, he or she would probably knock out a draft of a proposal or report without making notes or an outline.

Before you write, plan. Create a rough outline that spells out the contents and organization of your report or paper. The outline need not be formal. A simple list, doodles, or rough notes will do—use whatever form suits you.

Keep in mind that the outline is a tool to aid in organization, not a commandment etched in stone. If you want to change it as you go along, fine.

A big advantage of using an outline is that it helps you divide the writing project into many smaller, easier-to-handle pieces and parts. You might be intimidated by the thought of having to write the first draft of a 100-page user's manual. So use an outline to divide it into chapters, and write one at a time. That task is not nearly as overwhelming.

The organization of your outline depends on the type of document you're writing. In general, it's best to stick with standard formats. A laboratory report, for example, has an abstract, a table of contents, a summary, an introduction, a main body (theory, apparatus, procedures, results, and discussion), conclusions and recommendations, and nomenclature. A proposal will follow the outline of the RFP (request for proposal).

If the format isn't strictly defined by the type of document you are writing, select the organizational scheme that best fits the material. Some common formats are described in Table 1-2.

10. Interesting

"It is a sin to bore your fellow creatures," claims David Ogilvy, founder of Ogilvy & Mather, one of the world's largest advertising agencies. Ogilvy knows that a piece of writing must gain and keep the reader's attention if it has any hope of being read. Your technical reports compete with many other communications, such as letters, memos, trade journals, popular magazines, newspapers, newsletters, novels, the Sunday comics, radio, movies, videos, phone calls, and television. Be lively and lucid, not dull and boring. People in technical fields are human, too.

Table 1-2. Common formats for organizing technical material.

Order of location	An article on the planets of the solar system might begin with Mercury (the planet closest to the sun) and end with Pluto (the planet farthest from the sun).
Order of increasing difficulty	Computer manuals often start with the easiest material and, as the user masters basic principles, move on to more complex operations or functions.
Sequential order	Some technical activities, such as the installation of equipment, must be done in a certain order, and the presentation of instructions must follow that order.
Alphabetical order	This is a logical way to arrange a booklet on vitamins (A, B, B1, and so on) or a directory of company employees.
Chronological order	You can present events in the order in which they happened. History books are written this way. So are conference, trip, and call reports.

Problem/solution	The problem/solution format begins with "Here's what the problem was" and ends with "Here's how we solved it." This format is used frequently in case histories and user success stories.
Inverted pyramid	In the newspaper style of news reporting, the lead paragraph summarizes the story, and the following paragraphs present the facts in order of decreasing importance. You can use this format in journal articles, letters, memos, and reports.
Deductive order	Start with a generalization, then support it with facts, research results, examples, and illustrations. Scientists use this format in research papers that begin with the findings or main conclusions and then state the supporting evidence.
Inductive order	Start with one or more examples or stories, then lead the reader to the conclusion, idea, or principle that can be drawn from the examples. This is an excellent way to approach trade-journal feature stories.
List	Divide your discussion into a series of distinct points. Separate the points using subheads, bullets, or numbers. Present the points in order of priority.

2

How to Write Numbers, Units of Measure, Equations, and Symbols

Numbers, units of measure, mathematical equations, and symbols are used far more frequently in technical writing than in ordinary composition. Because numbers communicate much of the data in many technical documents, they must be written clearly and expressed consistently.

This sounds simple enough—until you remember there is more than one way to write a number. For example, does that transmitter in your communications system weigh three-quarters of a ton, ¾ ton, 0.75 ton, ⁷⁵⁄₁₀₀ ton, 1500 pounds, 1500 lb, or 1,500 lbs? Did it cost twelve and a half million dollars, 12.5 million dollars, $12½ million, $12.5 million, 12.5 × 10⁶ dollars, or $12,500,000? Can you write these numbers any way you please? Or is there a well-defined format for handling mathematics in writing?

As with grammar, spelling, and punctuation, there are set rules for writing numbers, units of measure, equations, and symbols. Some are rigid; others vary from one style manual to the next. In this chapter, we combine the most universally accepted rules with plain common sense; the result is the following set of guidelines for writing mathematical terms in a correct, consistent style.

Numbers

Lord Kelvin, creator of the Kelvin scale of temperature, once said, "When you can measure what you are speaking about, and express it in numbers, you know something about it." Unfortunately, few writers take the time to write numbers correctly, and more typos, errors, and inconsistencies occur in the mathematical portion of technical documents than in the prose portion. By sticking to a few simple rules, you can eliminate most of these mistakes.

Rule 1. Write out all numbers below 10.

The numbers one through nine are written out in English. You will find this rule the same no matter which grammar text or corporate style manual you consult. Most also recommend writing the number zero in word form rather than as a numeral. So you'd write:

> nine tractors
> one trial run
> zero quality defects
> five command centers

> The exception to this rule are numbers used with:

- units of measure
- age
- time
- dates
- page numbers
- percentages
- money
- proportions

Therefore, you would write:

2 yards
9-second delay
1 pound
6 years old
2 pm
October 7, 1992
page 3
4 percent
$3 (not $3.00)
70:1
70 to 1

In technical writing, any number *greater* than nine usually is written in numerals:

10 times better
6,000 abstracts
45,000 residents

Rule 2. When two or more numbers are presented in the same section of writing, write them as numerals.

Write all numbers as numerals when two or more numbers are presented in the same section of writing, as in this familiar Christmas ditty:

On the 12th day of Christmas, my true love sent to me: 12 drummers drumming, 11 lords a-leaping, 10 ladies dancing, 9 pipers piping, 8 maids a-milking, 7 swans a-swimming, 6 geese a-laying, 5 gold rings, 4 calling birds, 3 French hens, 2 turtle doves, and a partridge in a pear tree.

Writing numbers in a consistent format makes them easier to read and compare. It looks neater, too. Here is an example from a technical proposal:

The full-scale system contains 15 pumps, 5 fans, 5 ducts, and 3 heat exchangers.

There is one exception to this rule: If none of the numbers in the section is greater than nine, write them all out as words:

The pilot-plant system contains five pumps, one fan, one duct, and two heat exchangers.

Rule 3. Write large numbers in the form most familiar to your audience and easiest to understand.

There are many different ways to write a given number:

209,000,000
209 million
2.09×10^8
209×10^6
two hundred and nine million

Unfortunately, not all sources agree on how to handle these large numbers. We have adopted those rules that make for a clear, readable document.

Numbers in the thousands should be written with the separating comma; this makes them easier to understand at a glance:

14,948
199,492
51,000

While some style guides say to omit the comma in four-digit numbers, we think it should be left in; therefore, the numbers in the left column should be rewritten in the form shown in the right column:

3000	3,000
1200	1,200
1589	1,589

In Europe, a period is used in place of the comma. Thus, European technical writers would write 3,000 as 3.000 and 148,677 as 148.677. The number 2,344,000 would be written European style as 2.344.000.

Numbers in the millions can be written in two ways: as numerals or as a numeral plus the word *million*. Use the numeral-and-word format when the last five or six digits are all zero:

2 million instead of 2,000,000
3.5 million instead of 3,500,000
150 million instead of 150,000,000
15 million instead of 15,000,000
1.5 million instead of 1,500,000

Use numerals when the last five or six digits are other than zero:

5,936,999 instead of 5.936999 million
6,000,003 instead of 6.000003 million
1,234,567 instead of 1.234567 million

What about writing numbers in the billions? *Billion* means a thousand million in the United States but a million million in most other countries. Therefore, in technical documents that reach a multinational audience, numbers in the billions should be written as numerals; if the readership is strictly U.S. citizens, use the numeral-and-word format: 1.7 billion.

In writing amounts of money, use a numeral followed by the word *million* or *billion*. Thus you would write:

$67 million instead of $67,000,000
$6.7 billion instead of $6,700,000,000

Numbers above the billions are usually written in numerals, not as numerals and words, because the terms for these huge numbers—*trillion, quadrillion, quintillion*—are not familiar to the average reader:

1,500,000,000,000
69,584,294,668,234,576

Another method for writing numbers, scientific notation, expresses numbers using multiplication and exponents. In scientific notation, 1.5 million is written 1.5×10^6, which means 1.5 multiplied by 1,000,000. Other examples:

$1.5 \times 10^9 = 1,500,000,000$
$1.5 \times 10^8 = 150,000,000 = 150$ million
$10^3 = 1,000$
$7 \times 10^1 = 70$

Scientific notation is helpful when writing extremely large numbers that are difficult to read when expressed as straight numerals:

$1,000,000,000,000 = 10^{12}$
$602,300,000,000,000,000,000,000 = 6.023 \times 10^{23}$

As a rule, it is better to use numerals than scientific notation. Scientific notation may be compact, but its format is difficult to type and is likely to be reproduced incorrectly by graphic artists, typesetters, and desktop publishers. Besides, it's easier for the reader to think in terms of 100 million, 7.5 million, and 12,000 than 10^8, 7.5×10^6, and 1.2×10^4.

Rule 4. Place a hyphen between a number and unit of measure when they modify a noun.

Whenever a number and unit of measure are compounded to form an adjective, they are separated by a hyphen:

2-week-old culture
15,000-volt charge
8-pound baby
12-inch-long ruler

The hyphens make it clear that the numbers modify the units of measure rather than the nouns. For example, without hyphens, the reader might interpret the phrase *12 inch long ruler* as meaning you have 12 of something called "inch-long ruler." With hyphens, it's clear we're talking about rulers that are 12 inches long.

No hyphen is used between the number and unit of measure when they do *not* form a compound adjective. So in the sentence *The feed-pipe section goes on for 47 feet*, there is no hyphen because *feet* is used as a noun. There would be a hyphen, however, if we wrote about a *47-foot-long feed-pipe*.

Note that all numbers with units of measure are written as numerals, as we stated in rule 1.

Leaving out the hyphen between the number and unit of measure is one of the most common mistakes technical writers make. If you want your writing to be correct grammatically as well as technically, you must get in the habit of using it.

Rule 5. Use the singular when fractions and decimals of one or less are used as adjectives.

At first glance, saying *0.8 tons* may seem correct. But 0.8 is a decimal representation for eight-tenths, or four-fifths. Four-fifths is less than one, and when there's one or less of anything, it's singular, not plural. You wouldn't say "half a tons," would you?

The correct usage is *0.8 ton*—with *ton* being singular. Similarly, you would write:

0.9 pound
0.3 centimeter
0.44 cubic foot
0.5 cup of coffee
0.25 inch
0.8 kilometer
¼ mile

Rule 6. Write decimals and fractions as numerals, not words.

Decimals and fractions should be written as numerals; this form is more concise and readable than when they are written as words:

zero point three four	0.34
five point five eight nine	5.589

three-fifths   ⅗ or 0.6

four-tenths   ⁴⁄₁₀ or 0.4

We recommend writing fractions as decimals whenever possible. Standard typewriters have keys for only ½ and ¼; all other fractions must be constructed with the stroke or slash mark (/): 1/3, 1/5, 1/8, 3/16. Decimals are easier to type in a uniform fashion and are less likely to be reproduced incorrectly:

34/100   0.34

6/20   0.3

This is simple enough for fractions that convert cleanly to decimals; to get the decimal equivalent of ¼, for example, you divide 1 by 4 and get 0.25.

For fractions that do *not* convert cleanly to decimals—⅓, ⅔, ⅐—the quotient will often be a string of repeating digits. The string is written out once, and then a bar is placed over it to indicate that the string repeats infinitely.

For example, to convert ⅓ to decimals, we divide 1 by 3 and get 0.333333333 . . . the 3 repeats forever. This is represented by writing the repeating digit or string of digits once and putting a bar over it or them—thus the decimal equivalent of ⅓ is written $0.\overline{3}$. The bar indicates that $0.\overline{3}$ is shorthand notation for the number 0.3333333333 with the 3 endlessly repeating.

In the same way, the decimal equivalent of ⅔ is $0.\overline{6}$. To round off to 0.67 introduces a degree of inaccuracy; 0.67 is *not* precisely equal to ⅔.

Therefore you would convert fractions to decimals as follows:

½   0.5
¾   0.75
⅙ $0.1\overline{6}$
⅜   0.375
⅐ $0.\overline{142857}$
⅔ $0.\overline{6}$

If you haven't taken a math class lately, this bar notation may seem strange to you, but it is the correct way to express as decimals fractions with repeating quotients.

Rule 7. Treat decimal representations consistently, especially when presenting them in columns, rows, or groups.

A zero is always placed before the decimal point in numbers less than one:

.34   0.34

.5000   0.5000

The purpose is to assure consistency when such decimals are mixed with decimals greater than one, especially in columns or rows:

225.422
  0.34
  0.5000
  1.99
16.88

The question always arises as to the best way to line up numbers when listing them in a column. Our preference is to have the decimal points aligned vertically, as in the above example. We like this method best because it enables easy addition of the columns of numbers.

Other style guides may suggest lining up the numbers flush left:

225.422
0.34
0.5000
1.99
16.88

or flush right:

225.422
  0.34
  0.5000
   1.9
 16.88

Rule 8. Do not inflate the degree of accuracy by writing decimals with too many digits.

When you divide decimals, the quotient is written with the same number of digits after the decimal point as the longer of the numerator or denominator. (The numerator is the number appearing above the line in a fraction; the denominator is the number appearing below the line.)

Therefore, when you divide 2.34 by 5.55 and your calculator displays the answer 0.4216216, you should express this as 0.42, rounding off to the nearest hundredth. Writing it as 0.422 implies accuracy to the nearest thousandth, which is not the case (you can't have a quotient more accurate than the numbers being divided).

When rounding off, if the digit being dropped is five or greater, the digit being rounded is increased by one. Otherwise, the digit being rounded is unchanged. So 0.425 rounds off to 0.43, but 0.421 rounds off to 0.42.

Writing a number as a decimal implies precision to the last decimal place: *0.5 cup of coffee* indicates that the coffee is measured to the nearest tenth of a cup. If a number is merely an approximation (*half a cup of coffee* is roughly half a cup, more or less), do not give it an inflated degree of accuracy by writing it as a decimal.

Rule 9. If a number is an approximation, write it out.

If you measure ³⁄₁₆ of an inch with an ordinary ruler, your measurement is accurate to a sixteenth of an inch—and *not* to the tenthousandth of an inch, which would be implied if you were to write the fraction ³⁄₁₆ as its decimal equivalent, 0.1874. Whenever you write a number as a decimal, you are stating that the number is accurate to the last decimal place. By writing an approximation as a decimal, you imply a degree of precision that is not there. Therefore, if a number is an approximation, write it out:

half a glass of water
a quarter of a mile down the road

one-third less energy required
three-quarters finished
twice the pressure

You can use "hedge" words and phrases such as *about, up to, almost, more or less, roughly, approximately,* and *on the order of* to indicate an even greater degree of uncertainty or approximation in your numbers:

about half a glass of water
almost a quarter of a mile down the road
approximately one-third less energy required
roughly three-quarters finished
up to twice the pressure

Rule 10. Spell out one of two numbers—usually the shorter—that appear consecutively in a phrase.

If all of the numbers in a phrase are written as numerals, it can confuse the reader. For example, when you write about *12 21-module software packages,* the reader might think you accidentally left a space and meant to write *1,221-module software packages.* Or, the reader might assume "12 21" is the part number or code of the software module.

The solution is to write out one of the numbers—preferably the shorter of the two—in word form. This way, the reader who sees *twelve 21-module software packages* knows you're referring to a quantity of twelve software packages containing 21 modules each.

Here are a few more examples; the phrases in the left-hand column should be rewritten as shown at right:

four four-color photos	four 4-color photos
2 3-inch wrenches	two 3-inch wrenches
11 60-ohm resistors	eleven 60-ohm resistors
three five-person teams	three 5-person teams
45 4,500-component radar systems	forty-five 4,500-component radar systems

When one of the numbers is modifying a unit of measure, as the number three is in *two 3-inch wrenches,* that number is written in numerals and the other number in the phrase is spelled out, regardless of which is shorter:

two three-inch wrenches	two 3-inch wrenches
52 20-foot fuel rods	fifty-two 20-foot fuel rods

When two numbers in two different clauses are separated only by a comma, write one as numerals and the other as words:

When the count reaches 6,420, 420 units will be recalled.	When the count reaches 6,420, four hundred and twenty units will be recalled.

Rule 11. Do not begin a sentence with numerals.

Readers, like drivers, become used to certain traffic signals. Periods, exclamation points, and question marks are red lights that signal the end of a sentence. A word with the first letter capitalized is the green light that signals the start of a sentence. Therefore, you should avoid starting a sentence with anything other than a capital letter—and that includes symbols and numerals.

Although it is sometimes difficult to avoid, never begin a sentence with a numeral: It confuses the reader. You can either write the number out in word form or, if the spelled-out form is unwieldy, rewrite the sentence so that it no longer begins with a number:

2,000 test subjects participated in the experiment.	Two thousand test subjects participated in the experiment.
$250,000 went into the preparation of their bid for the new defense system.	A quarter of a million dollars went into the preparation of their bid for the new defense system.

154,612 solders will go
through the training program
in the next 10 years.

In the next 10 years, 154,612
soldiers will go through the
training program.

Units of Measure

Rule 12. Keep all units of measure consistent.

Units of measure must be consistent. Consistency allows the reader to make easy comparisons between numbers. For example, if you are discussing temperature, choose one standard of measurement and stick with it. Using degrees Centigrade in one place, degrees Fahrenheit elsewhere, and degrees Kelvin in other places is confusing and therefore inappropriate.

When you're inconsistent in your use of units of measure, you force the reader to do extra work converting sets of numbers to like units. Making readers do your work wastes their time and also creates opportunity for error: Readers may make a mistake or not even know how to calculate the conversion.

The best advice is to pick a set of units and stick with it throughout your text, tables, and illustrations. A common mistake is to write text using one set of units, then take a table or chart from another source that contains other units of measure and drop it in without doing the conversion. Avoid this.

Rule 13. Use the correct units for the system of measurement you have chosen.

Length, mass, volume, pressure, and other physical characteristics can be expressed in two basic systems of units: English and metric. Many U.S. scientists and engineers prefer the English system, which measures length in feet and inches, mass in slugs, force in pounds, time in seconds, and pressure in pounds per square inch. Europeans, Asians, and almost everyone else favor the metric system.

There are two different versions of the metric system. The older version is the centimeter-gram-second (cgs) system. As the name implies, it measures length in centimeters, mass in grams, and time in seconds.

The modern version of the metric system is the Système International (SI). In SI, length is measured in meters, mass in kilograms, force in newtons, time in seconds, and pressure in pascals.

Table 2-1 shows the basic units of measure in each system. As you can see, the slug and kilogram are units of mass; the pound and newton, units of force.

Metric has become an international standard; the United States is one of only a handful of countries still using the English system.

Of the two versions of the metric system, SI is more widely used than cgs. If you are submitting a paper for publication in a technical journal or for presentation at a professional meeting, the author's guidelines supplied by the publisher or conference sponsor will probably specify which system of measure to use in your paper.

Be especially careful using units of measure for weight and mass when writing for lay readers. The average person is not familiar with the slug as a unit of measure and does not clearly understand the difference between weight and mass.

Table 2-1. Units of measure for English, cgs metric, and SI metric systems.

Characteristic	English unit	cgs unit	SI unit
time	second	second	second
mass	slug	gram	kilogram
force	pound	dyne	newton
length	foot	centimeter	meter
pressure	pound per square inch	dyne per square centimeter	pascal
energy	foot-pound	erg	joule

As you may recall from high school physics, force, according to Newton's second law of motion, is mass multiplied by acceleration. Weight, which is the force gravity exerts upon a body, is equal to the mass of the body times the local gravitational acceleration.

Strictly speaking, then, pounds and kilograms are not interchangeable, since the former is force and the latter is mass. However, under standard conditions of gravity on Earth, 0.4536 kilogram always weighs 1 pound, and both the pound and the kilogram are commonly—if not accurately—used as measures of weight. The slug, accepted in the scientific community, is practically unheard of in everyday usage.

Rule 14. Write basic units of measure in word form, derived units of measure as symbols.

Most units of measure can be written in one of three ways: as a word, as a symbol, or as an abbreviation. For example, you can express the unit of measure of time as a word (*seconds*), use a symbol for it (the proper symbol for second is *s*), or abbreviate it as *sec*.

Other examples:

Unit of measure	Symbol	Abbreviation
inch	"	in.
ampere	A	amp
kilogram	kg	kilo
pound	#	lb.

Which form should you use? It depends on where and how the unit of measure is referred to in your document.

For the basic units of measure—those involving measurement of a single quantity, such as time, length, or height—write out the unit as a word:

12-inch ruler
6-second delay
48 pounds

24 hours
90-ohm resistor
28 meters

It is better to use too few than too many abbreviations and symbols. And, despite their official status, symbols for units of measure such as seconds, hours, meters, and inches (s, h, m, in.) look strange and unfamiliar to many readers. Therefore, if the unit of measure is short and can be written simply, write it out as a word.

For *derived* units of measure—those formed by multiplying and dividing other units (miles per hour, pounds per square inch)—write the units as symbols:

32 ft/s^2
667 m^3/s
500 J(kg.K)

We use symbols here because writing out compound units of measure in word form becomes cumbersome: It's cleaner and simpler to express radiance as $W/(m^2.sr)$ than as *watt per square meter-steradian.*

There are some derived units that have two symbols—one being an abbreviation of the name of the unit, the second showing how the unit is derived. For example, power expressed in watts can be written using *W*, the symbol for the word *watt*, or *J/s* for joule per second.

Our recommendation is that when you are writing these special derived units that have a symbol representing the word as well as one representing the derivation, use the one that represents the word. The reason is these are the forms most familiar to your readers. Thus you would write Hz for hertz, J for joule, and V for volt. Other examples:

Unit	*Symbol*
watt	W
megawatt	MW
coulomb	C
farad	F

ohm	Ω
tesla	T
henry	H

Rule 15. Indicate multiplication of units by a raised dot ($\cdot$), division by a slash, (/).

To indicate the multiplication of units written as symbols, use a raised dot ($\cdot$), not a multiplication sign ($\times$) or hyphen (-). Indicate division with a slash, also known as a stroke or solidus (/):

0.3 J per mol $\times$ K	0.3 J/(mol $\cdot$ K)
32 ft per s^2	32 ft/s^2
10 kg per m-s^2	10 kg/(m $\cdot$ s^2)

If, however, you write out a unit in words, not symbols, indicate multiplication with a hyphen and division with the word *per:*

Top speed is 60 miles per hour.
The force was 240 kilogram-meters per second squared.
Typical plants process 80 million cubic feet of fly ash per day.

Rule 16. Write secondary units in parentheses after the primary units.

Writers who choose metric as their primary system of measurement may want to give English equivalents for readers more comfortable with inches and pounds than centimeters and kilograms.

The opposite is also true: American scientists and engineers preparing papers using English units may also want to give metric equivalents for European readers.

When doing so, the secondary unit of measure appears in parentheses immediately following the primary unit of measure:

We used a 10-meter (32.8-foot) section of 5.08-centimeter (2-inch) diameter pipe.
The operating range is from 10° C (50° F) to 65° C (149° F).

Consistency demands that if you convert primary units to secondary units and display them in parentheses, you should do so throughout the document wherever units of measure are used, not just sporadically or at random.

Equations

Mathematical equations frequently communicate more effectively and more eloquently than words; the most well-known scientific thought in Western civilization is probably $E = mc^2$.

To writers, editors, proofreaders, typists, typesetters, and graphic artists, however, these cumbersome, complicated formulas are a production nightmare.

To begin with, standard typewriters and word-processing software cannot make the superscripts, subscripts, brackets, arrows, mathematical signs, Greek letters, and other symbols used in equations.

Worse, these strange-looking marks can indeed be "Greek" to the people involved in editing and producing the manuscript. Their lack of familiarity with the author's specialized notation often results in equations that are incorrectly typed and reproduced in the final document. And as you know, even something as innocent as a misplaced subscript or exponent can completely change the meaning of a mathematical expression.

Fortunately, with a little extra care and patience, these mistakes can be eliminated.

Do not type equations on your first draft—write them in longhand as clearly as possible. Trying to improvise on a standard word-processing system will only frustrate and confuse both author and editor. If you use desktop publishing, however, your software may allow you to key some simpler equations directly into your document. (Some mathematics programs, such as Mathematica™ and Maple V™, display and print out publication-quality equations and calculations in standard math notation.)

In its final form, the equation will be typed, typeset, or penned neatly by hand. It's essential that authors proofread their own work to make sure the equations are stated correctly before their manuscripts are reproduced and distributed; a proofreader is not a mathematician or engineer and will probably not be able to spot most equation errors.

Equations will never be fun to write. But by following the hints and the guidelines below, you can handle them properly and with a minimum of personal agony.

Rule 17. Use too few rather than too many equations.

While many scientists and engineers tend to communicate using equations rather than words, you should nonetheless keep the number of equations in your document to a minimum. The reason, already stated, is the potential for error in the writing and reproduction of equations in technical documents, as well as the difficulties of typing and typesetting them.

If you are writing for a nontechnical audience, strive to eliminate equations altogether. While equations may be second nature to you, many people fear or are utterly bored by mathematics, and seeing an equation is a turn-off that may prevent them from reading or even picking up your article, booklet, or report.

Rule 18. Center and number equations on a separate line in your document unless they are short and simple.

Most equations are lengthy and unwieldy—too big to treat as a regular part of the sentence or handle in paragraph form.

The solution is to write each equation on a separate line:

The general first-order linear equation is

$$dy/dx = p(x)y + q(x) \tag{1}$$

and the general second-order equation is

$$d^2y/dx^2 = p(x)dy/dx + q(x)y + r(x). \tag{2}$$

Note that the equations are numbered in the order in which they appear. This makes it easy to refer to them later in the text:

We can rewrite the general first-order linear equation (Eq. 1) as

$$\frac{dy}{dx} - p(x)y = q(x). \qquad (3)$$

Centering and numbering equations makes the document neater and more readable, and enables you to refer to equations by number.

Equations that are short and simple may be placed on a separate line or run into the text—whichever you prefer:

On Ludwig Boltzmann's gravestone is carved his formula for entropy:

$$S = K \log W. \qquad (1)$$

On Ludwig Boltzmann's gravestone is carved his formula for entropy, $S = K \log W$. Boltzmann committed suicide in 1906 at the age of 62.

Rule 19. Keep all equal signs, division lines, fraction lines, multiplication signs, plus signs, and minus signs on the same level.

To make sure your equations have a consistent format and don't wander all over the page like a sine wave, keep all plus, minus, multiplication, and division signs aligned horizontally with the equal sign:

$$\Delta P_t = \frac{4}{\pi} Ne\mathrm{Re}_D \frac{nV}{D^4} \left(\frac{L}{D}\right)$$

$$A = B + \frac{C}{D} + \frac{E}{F}$$

$$x = e^{-bt}(C_1 \cos \theta + C_2 \sin \phi) + \frac{F_0}{K_1}$$

$$\frac{dy}{dx} = p(x)y + q(x)$$

For a series of equations, write them so their equal signs are aligned vertically:

$$x = A + B + C$$
$$y = D + E$$
$$z = F + G$$

Rule 20. Punctuate words used to introduce equations just as you would words forming part of any sentence.

Like any other phrase, an equation is part of your sentence. Why, then, do writers constantly introduce formulas with punctuation that does not fit the sentence structure? A high school physics test, for example, contains this sentence:

The current in the wire is calculated using Ohm's law.

This sentence is properly punctuated. Adding a colon serves no purpose and is unnecessary:

The current in the wire is calculated using: Ohm's law.

The construction should not change if Ohm's law is written as an equation. Yet many authors would write the following:

The current in the wire is calculated using:
$$E = IR$$

There is no logical reason for using punctuation incorrectly where equations appear. Punctuate the words introducing the equation as if the equation were just a regular part of the sentence—which it is. Therefore, in our example you would write:

The current in the wire is calculated using
$$E = IR.$$

Note that there is a period after the "R" in $E = IR$. The rule is to put a period after the last term in the equation if the equation ends the sentence and is followed by a new sentence:

The current in the wire is calculated using
$$E = IR.$$
Here E is current, I is electric potential, and R is resistance.

If the equation is in the middle part of the sentence and not the end of the sentence, no period is placed after the equation:

The current in the wire is calculated using
$$E = IR$$
where E is current, I is electric potential, and R is resistance.

Many writers and editors unfamiliar with the language of mathematics seem to be intimidated by equations and are unable to correctly punctuate sentences in which equations appear. If you fit this category, it might help you to mentally substitute the phrase *this equation* for the actual formula when you punctuate the sentence. The sentence structure will be the same, but thinking in terms of words instead of numbers and symbols should make things easier:

The current in the wire is calculated by using this equation.
The current in the wire is calculated using this equation where E is current, I is electric potential, and R is resistance.

You then go back and insert the actual equation ($E = IR$) as a replacement for "this equation." When you do, do not change the punctuation you've chosen.

Symbols

Rule 21. Use too few rather than too many symbols.

Symbols provide a way of writing units of measure, mathematical variables, physical constants, scientific names, and other terms in an abbreviated fashion. They can reduce writing and word-processing time and make sentences briefer. A few examples of words and their symbols are presented in Table 2-2.

Some of these symbols (%, +, $, #) are probably familiar to you; others (☿, ✳, ↑) you may not recognize. You will never write for an audience that knows the meanings of all of the hundreds of symbols used in technical literature. Therefore, if there is the slightest chance that any part of your audience will not recognize a symbol, write out the word instead. Avoid confusion and misunderstanding by using too few rather than two many symbols.

Even readers in your specialized field may not know all the symbols you do. If you are a chemist writing for other chemists, it's a safe bet that writing H_2SO_4 instead of *sulfuric acid* will not confuse anybody. But

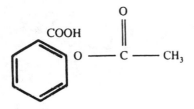

might give some readers a headache. The writer would be better off simply writing out the name of the compound—in this case, acetylsalicylic acid (aspirin).

When they are properly used, symbols, like other abbreviations, can make your sentences shorter and more readable. Especially in equations, substituting symbols for words makes the writing concise:

force equals mass times acceleration	$F = ma$
energy equals mass times the speed of light squared	$E = mc^2$
the sum of the squares of the lengths of the sides of a right triangle is equal to the square of the length of the hypotenuse	$a^2 + b^2 = c^2$

Rule 22. Define the symbols you use.

Make sure your reader knows what you are talking about by defining the symbols you use. Clearly identify each symbol as it is introduced in the text. The reader should not have to refer to

Table 2-2. Symbols.

Word or concept	Symbol
dollars	$
number	#
percent, percentage	%
and	&
degree	°
positive	+
negative	−
less than	<
greater than	>
mercury (the element)	Hg
Mercury (the planet)	☿
snow	✳
gas	↑
ohm	Ω
square root	√
cent	¢
Pisces (zodiac)	⊁
Silurian soil	S
copyright	©
radiation	X

an appendix, glossary, or technical dictionary to understand your meaning:

The most acidic compound known is $HClO_4$ (perchloric acid). To convert Celsius to Fahrenheit, use

$$°F = 1.8(°C) + 32°$$

where F is degrees Fahrenheit and C is degrees Celsius.

In addition to defining symbols as you introduce them in your text, if you use four or more symbols you should also include a nomenclature section at the end of your document. The nomenclature section lists, in alphabetical order, the symbols used, along with their definitions and units of measure, if any.

This is extremely helpful for readers going through the calculations in your paper or report.

Rule 23. Avoid duplication of symbols.

Electric resistance, degrees Rankine, the universal gas constant, and the mean radius are all symbolized by R. In the same way, T stands for the prefix *tera* (10^{12}), for temperature, and for the tesla, a unit of magnetic flux.

A given symbol frequently stands for more than one variable or unit of measure. This is because there are not enough symbols to go around, and technical experts in one discipline will adopt a symbol regardless of whether it's already being used in another area of science.

You may find yourself using several terms, all represented by a single symbol. Obviously, writing these terms as symbols must be avoided. Either write out all terms as words or redefine them with new, nonstandard symbols:

The R in the test wire was 60 ohms; ambient temperature was 520° R.	The resistance in the test wire was 60 ohms; ambient temperature was 520° R.
	The r in the test wire was 60 ohms and ambient temperature was 520° R where r is resistance and R degrees Rankine.

Table 2-3. Sample nomenclature section.

Symbol	Definition	Unit
B	magnetic inductance	weber per meter squared
C	capacitance	farad
E	electric intensity	volt per meter
L	inductance	henry
R	resistance	ohm

Rule 24. Fit symbols grammatically into the structure of your sentence.

Philosopher J. J. C. Smart wrote:

If the argument is valid, that is, if r really does follow from p and q, the argument will lead to agreement about r provided that there is already agreement about p and q.*

Philosophy and logic aside, note that Smart has fit his symbols (r = conclusion, p = first premise, q = second premise) neatly into the structure of his writing. The sentence, though wordy, is properly punctuated.

Symbols are not meaningless, alien markings; they are substitutes for words. And so, symbols must be handled with the proper rules of grammar and punctuation, just as you would handle the words they represent.

---

*George I. Mavrodes, ed., *The Rationality of Belief in God* (Englewood Cliffs, N.J.: Prentice Hall, 1970), p. 97.

3

A Few Useful Rules of Punctuation, Grammar, Abbreviation, and Capitalization

There are hundreds of rules of grammar and punctuation; in this chapter, we present those that come up most often in technical writing.

Punctuation

Rule 25. Hyphenate two words compounded to form an adjective modifier.

In the sentence *This is a sure-fire tip,* the two words *sure* and *fire* form an adjective compound modifying *tip.* When two or more words are compounded to form an adjective that precedes a noun, they are usually hyphenated:

space time continuum	space-time continuum
state of the art technology	state-of-the-art technology
long range, high power radar	long-range, high-power radar

Hyphens are used in compound modifiers because they help the reader avoid confusion. For example, if a compound such as *first-order reaction* were not hyphenated, the reader might expect *first* to modify the phrase *order reaction;* when the hyphen is included, the reader sees a reaction of the first order.

No hyphen is needed following adverbs that end in *-ly:*

technically accurate manual
artificially induced sleep
financially stable organization

However, adjectives ending in *-ly* are hyphenated when they are used with present participles, as in *friendly-sounding voice.*

Many compound words that are hyphenated *before* a noun are not hyphenated when they occur *after* the noun:

this is my up-to-date report.	Bring them up to date.
It was a well-known principle.	The principle was well known.

When a phrase like *up to date* follows the noun, the relationships between words are clear enough without hyphens so that readers will not be confused.

Do not hyphenate scientific terms, chemicals, diseases, and plant and animal names used as modifiers if no hyphen appears in their original form:

sulfur dioxide emissions
swine flu epidemic
an apple tree grove

Rule 26. Hyphenate two adjacent nouns if they are both necessary to express a single idea.

In English, the trend has been first to join frequently used compound nouns with a hyphen and eventually to make them a single word:

air craft	air-craft	aircraft
type setting	type-setting	typesetting

screen play	screen-play	screenplay
data base	data-base	database

The hyphen is used to make the connection between the two words immediately obvious to the reader. For example, the unhyphenated combination *feed pipe* seems, at first reading, to request food for a hungry pipe. With the hyphen, we instantly recognize the compound *feed-pipe* as the pipe through which fluids flow into a process system. Some other examples of how a hyphen clarifies the meaning of a compound are shown below:

life form	life-form
tie rod	tie-rod
light year	light-year

The dictionary is the final authority on whether two words are separate, hyphenated, or joined into one word.

Rule 27. In a series of three or more terms with a single conjunction, use a comma after each except the last.

The comma after the next-to-last term (the term that comes just before the conjunction) is known as the serial comma. In *bacon, eggs, and juice,* the serial comma is the comma appearing after the word *eggs.*

Today, writing authorities are divided as to whether the serial comma should be used. The major reference works on English grammar recommend using this comma. We agree. It can add clarity and make sentences flow more smoothly. Therefore, you should write:

The four most abundant elements in the earth's crust are oxygen, silicon, aluminum, and iron.

Reports, proposals, and manuals are the responsibility of the technical-writing department. The advertising department handles brochures, catalogs, and press kits.

She is a communications consultant for CNN, Fox, NBC, and CBS.

At the convention, we listened to the early music of Neil Diamond, Gladys Knight and the Pips, and Bruce Springsteen.

Rule 28. Omit the period at the end of a parenthetical expression within a sentence; retain it if the entire parenthetical expression stands alone as a sentence.

In technical writing, sentences often end with an expression within parentheses. Punctuate the enclosed material in the usual way but omit the final period in the parenthetical expression. Punctuate the sentence outside the parentheses exactly as if the parenthetical expression were not there:

Incorrect	*Correct*
The growth rate has increased by 12 percent each year (see Fig. 8.).	The growth rate has increased by 12 percent each year (see Fig. 8).
Now it's time to discuss even numbers (e.g., 2, 4, 6.).	Now it's time to discuss even numbers (e.g., 2, 4, 6).

The rule also applies to parenthetical expressions that appear in the middle of a sentence, unless it is a question mark or an exclamation point:

It is extremely unlikely that two people (barring identical twins) share precisely the same gene combinations.

At the science fair, the chemist (or was he more of an alchemist?) turned red wine into milk before our eyes.

If the entire parenthetical expression is a complete sentence in itself, and not contained within another sentence, the period at the end of the sentence goes inside the parenthesis.

(We will discuss software below.)

Grammar

Rule 29. Avoid dangling participles.

A verb ending in -*ing* is called a present participle. Grammarians say a participle is *dangling* when it is attached to the wrong subject in a sentence. Consider this example:

Turning over our papers, the chemistry examination began.

This sentence says that the chemistry examination began after it turned over our papers! Here, the participle *turning* is dangling because it is attached to the wrong subject: the chemistry examination. It should refer to the implied *we* that did the actual turning over of the papers. The correct form is:

Turning over our papers, we began the chemistry exam.

Danging participles occur frequently in technical writing, sometimes resulting in absurd sentences. Sentences with dangling participles must be rewritten to make their meanings clear.

King Tut's tomb was unearthed while digging for artifacts.	Archaeologists unearthed King Tut's tomb while they were digging for artifacts.
Sitting serenely in the laboratory, the apple tasted great to her.	Sitting serenely in the laboratory, she enjoyed the great taste of the apple.

Rule 30. Avoid run-on sentences.

When a comma or no punctuation is used between two complete sentences, the result is referred to as a "run-on" sentence. A run-on sentence loses the reader in an unbroken string of thoughts, leaving him or her breathless and confused.

The vacuum tube burned out, it will be replaced.

The computer is down, you must fix it soon.

These sentences are punctuated incorrectly. When two or more independent clauses not joined by a conjunction (*and, but, for, or, when, as, because,* or *though*) are to form a single sentence, they must be separated by a semicolon or period:

The vacuum tube burned out; it will be replaced.

The computer is down. You must fix it soon.

Or, you can join the independent clauses with a comma followed by a conjunction:

The vacuum tube burned out, but it will be replaced.

The computer is down, and you must fix it soon.

Run-on sentences can muddle your meaning, and their use shows poor knowledge of punctuation. Adding the proper punctuation or forming several shorter sentences makes your writing clear and correct.

Rule 31. Avoid sentence fragments.

A complete declarative sentence must contain a subject and a verb. A sentence missing one of these essential elements is an incomplete sentence, or *sentence fragment:*

As we said last Friday during our meeting.

All of the people who wrote the report during the meeting.

Sentence fragments can be so long you hardly realize they're missing something until you come to the end:

Maxwell's remarkable discovery that the speed of propagation of electromagnetic effects is precisely the same as the speed of light in the same medium.

Well, what *about* Maxwell's discovery? The lack of a verb makes this an incomplete sentence and renders it frustrating to readers as well. But add a simple verb, and the words suddenly make sense:

Maxwell made the remarkable discovery that the speed of propagation of electromagnetic effects is precisely the same as the speed of light in the same medium.

Sometimes, skilled writers use very short sentence fragments deliberately to achieve added impact:

At Venture Electronics, we are fast. And reliable.

The new fluidized-bed combustion boiler should increase heat-transfer efficiency by up to 500 percent. And reduce energy costs by 50 percent.

Sentence fragments should be used sparingly. Too many departures from standard written English make readers uneasy and give the impression that the writer is uncertain of the rules of grammar.

Abbreviation

Rule 32. Spell out abbreviations at their first appearance, and use too few rather than too many.

Abbreviations proliferate in our technical society. People are as fond of the shorthand of abbreviations as they are of jargon, and it's no wonder that two of the most famous characters from the film *Star Wars* are known simply as R2D2 and C3PO.

Everything would be fine if we could guarantee that our readers would understand all the abbreviations we use, *but do not assume that they do*. You may know what OPEC, NASA, GE, and GNP stand for, but not everyone does. A college student may know what MIT, RPI, and UVM stand for, but these initials do not communicate meaning to everyone. Even the often-used TGIF (Thank God It's Friday) isn't universal.

When we rely on abbreviations, we start a process of "inbreeding" that may prevent us from clearly communicating our thoughts to people outside our discipline or department. Government employees may understand that DOT means Department of Transportation or that OMH means Office of Mental Health, but they had better not rely on these abbreviations if they need to explain these departments to an outsider. In the same way, bureaucrats or scientists may be able to roll off abbreviations such as MRV, HUD, DOT, EPA, LEM, and OAS, but they should not assume that laypeople will understand the meaning unless the words are spelled out first.

Explain every abbreviation you use, except when you're reasonably certain your audience understands them. The first time you use an abbreviation, write out the word, followed by the abbreviation in parentheses. In this way, you can be sure that all your readers are familiar with the abbreviation.

The integrated computer-aided system (CAS) is not a novelty. One major manufacturer of electronic systems has used a CAS approach on all its major projects since 1968.

There are exceptions. Some abbreviations have become so widely accepted that the abbreviation is used in place of the word itself. Examples include LSD, AIDS, and DNA. In these cases, it's better to use the abbreviation than spell the word out, since the spelled-out word may be meaningless to the reader (e.g., LSD is lysergic acid diethylamide).

A word to systems professionals: Perhaps to a greater degree than any other technical field, the world of computers is filled with special terms, abbreviations, and acronyms. Some acronyms are so widely used that systems people may have forgotten what the terms stand for.

The experienced writer learns to identify those acronyms that might not be recognizable to readers at a glance. You can't assume that nontechnicians are used to seeing the many acronyms that systems people see every day; even common acronyms like DOS (Disk Operating System) and CPU (Central Processing Unit) need to be written out the first time they are used. After spelling out what an acronym stands for, you may then use it throughout the rest of the document. This principle applies to the following common acronyms as well as to any others in the systems field:

ASCII   American Standard Code for Information Interchange

CASE   Computer-Aided Software Engineering

CIM   Computer-Integrated Manufacturing

CRT   Cathode-Ray Tube

EDI	Electronic Data Interchange
GUI	Graphical User Interface
KB	Kilobyte
MB	Megabyte
MIPS	Millions of Instructions per Second
RAM	Random-Access Memory
ROM	Read-Only Memory

Rule 33. Omit internal and terminal punctuation in abbreviations.

This rule is mainly a matter of style and taste. Some organizations tell their editors to punctuate abbreviations; others prefer to delete the little dots. We choose to omit internal and terminal punctuation because this style has a cleaner, more readable appearance. Thus we write 5 pm instead of 5 p.m. and write psi instead of p.s.i. Additional examples are:

R.S.V.P.	RSVP
U.F.O.	UFO
A.S.A.P.	ASAP

There is one major exception to this rule: The final period *is* used when the abbreviation spells out a single word:

no. (number)
fig. (figure)

But if the word is a unit of measure, the final period is omitted:

lb (pound)
yd (yard)
mi (mile)

Internal punctuation should be retained in proper names. Write J. J. Thompson, not J J Thompson.

Rule 34. The abbreviation for a specific word or phrase takes the same case (upper case or lower case) as the word or phrase.

An abbreviation for a single word begins in the same case as the word itself:

Prof. Morgenroth (Professor Morgenroth)
Rep. Tom Perkins (Representative Tom Perkins)
ed. (edition)
Westinghouse Corp. (Westinghouse Corporation)

When an abbreviation is formed from the first or first few letters of each major word of a compound term, it is called an acronym. Thus ATC is an acronym for air traffic control, VCR for videocassete recorder, and CD for compact disk. Usually acronyms are written in uppercase letters, regardless of whether the words they stand for begin with capitals.

VAR (value-added reseller)
VHF (very high frequency)
USA (United States of America)
OEM (original equipment manufacturer)
AM (amplitude modification)

However, acronyms that stand for units of measure are written in lowercase letters.

ppm (parts per million)
rpm (revolutions per minute)
mph (miles per hour)
bps (bits per second)

Rule 35. Avoid using signs in writing (" for inch, ' for foot), except when expressing information in tables.

Word processors make life easy for the technical writer—sometimes *too* easy. It's tempting to press the key for double quote marks instead of writing out the word *inches*. Likewise, the apostrophe is a convenient stand-in for the word *foot*.

But this convenience invites confusion. In regular prose, a minus sign could be mistaken for a misplaced hyphen, dash, or underscore, while the sign would be clear if used in a table. The double quote marks immediately program readers for a quotation, not a measurement in feet. Even the symbol for *at* (@) can lead to confusion. So, simply avoid all unnecessary signs. When you write out the word, you opt for total clarity:

"A 15" opening in the wall allows proper ventilation," said Dr. Jones.	"A 15-inch opening in the wall allows proper ventilation," said Dr. Jones.
Standard container hold 75# of #5 carbon black.	Standard containers hold 75 pounds of no. 5 carbon black.

Capitalization

Rule 36. Capitalize trade names.

Every so often, a trademarked product comes into general use and people begin to use the word generically. When you see the word *Kleenex* instead of tissue, be aware that a trade name is being used and must therefore begin with a capital letter. Some product names have become English words in their own right, and they do not require a capital. For example, *fiberglass* is a word derived from Fiberglas.

Technical writers, as we have so often said, must uphold a high standard of accuracy. Don't write *I need to xerox this report.* Xerox is the name of a company and of one copier; it is not a synonym for the word *photocopy.*

To help you keep in mind some of the current trade names often confused with generic products or processes, here is a list of some that are, in a way, dangerously familiar.

Trade Name	Standard-item Name or Approximation
Astroturf	artificial grass
Band-Aid	bandage
Bufferin	buffered aspirin
Crescent Wrench	crescent-headed wrench

Electromatic	electric typewriter
Formica	laminated plastic
Frigidaire	refrigerator
Frisbee	flying disk
Highlighter	yellow marking pen
Jacuzzi	whirlpool bath
Jell-O	gelatin dessert
Kitty Litter	cat litter
Kodak	camera
Liquid Paper	correction fluid
Magic Marker	permanent marker
Mace	chemical repellent
Multigraph	multilith press
Novocain	painkiller
Omnifax	facsimile system
Ping-Pong	table tennis
Plexiglas	clear acrylic plastic
Polaroid	instant camera
Realtor	real-estate agent
Rolodex	rotating card file
Scotch Tape	clear tape
Selsyn, Autosyn	synchro
Styrofoam	extruded plastic
Sweet 'n Low	sugar substitute
Tabasco	red-pepper sauce
Univac	computer

Valium	muscle relaxer
Vaseline	petroleum jelly
Velcro	fabric fastener
WaterPic	water-jet teeth cleaner
Windbreaker	waterproof jacket
Winnebago	recreation vehicle
Wite-Out	correction fluid
Xerox	photocopy

Companies that have taken out a trademark on a product want to protect the use of its name. *Sanka* is advertised as *Sanka brand coffee* because its manufacturer does not want to encourage the public to think of Sanka as being synonymous with other competitive decaffeinated coffees. At one time, *aspirin, borax, linoleum, dry ice, shredded wheat,* and *trampoline* were trade names, but today they have lost all association with their original manufacturers.

Rule 37. Do not capitalize words to emphasize their importance.

Some writers capitalize certain words to make them important or stand out. Unfortunately this tendency—we call it "cap fever," the uncontrollable urge to emphasize words by capitalizing them—cheapens those words that must be capitalized and gives writing an undesirable, antiquated flavor:

Advertising and publicity can enhance the Value Package of your product.	Advertising and publicity can enhance the value package of your product.
Certain kinds of fuels can cause Fuel Starvation as cells age.	Certain kinds of fuels can cause fuel starvation as cells age.

Burning is a Chemical Reaction in which oxygen atoms combine with the atoms of the substance being burned.	Burning is a chemical reaction in which oxygen atoms combine with the atoms of the substance being burned.

There are possible exceptions to this rule. A graphic designer may choose to use a nonstandard style of capitalization for headlines, subheads, titles, and captions. Although it is technically uncalled for, this gambit may lend a more pleasing, eye-catching appearance to a brochure, advertisement, magazine, or book jacket. For maximum clarity, do not abuse the rules of capitalization for an effect that might cause your reader any puzzlement or might strike the reader as odd or artificial.

Rule 38. Capitalize the full names of government agencies, companies, departments, divisions, and organizations.

Official names and titles are capitalized:

Air Pollution Control Division
U.S. Small Business Administration
Spartan Engineering Company

Do *not* capitalize words such as *government, federal, agency, department, division, administration, group, company, research and development, engineering, manufacturing,* and *quality control* when they stand alone. They are capitalized only when they are part of an official name:

This is a problem for Research and Development, not Engineering.	This is a problem for research and development, not engineering.
	This is a problem for the Research and Development Department, not the Engineering Department.
She is the head of her Division in the Company.	She is the head of her division in the company.

Rule 39. Capitalize all proper nouns unless usage has made them so familiar that they are no longer associated with the original name.

Science offers many rewards, one of which is the pleasure of having your discovery named after you. Hundreds of theories, laws, formulas, numbers, and units of measure have been named for the scientists who conceived them. These words retain the capital letters used in the proper names from which they derive:

Kelvin scale (Lord Kelvin)
Heisenberg uncertainty principle (Werner Heisenberg)
Mach number (Ernst Mach)

Repeated usage has made the more common proper nouns and adjectives so familiar that most people no longer associate them with the names of their founders. When this happens, the capital letter is dropped.

diesel (Rudolf Diesel)
ampere (Andre-Marie Ampère)
hertz (Heinrich Rudolf Hertz)
ohm (George Simon Ohm)

4

Principles of
Technical Communication

Making your writing readable requires more than just an awareness of spelling, grammar, and syntax. It means developing a coherent, precise style. There are several valuable principles that guide technical writers in forging a lively, concise, and individual way of expressing ideas. Here are a few of the key principles of technical composition.

Rule 40. Use the active voice.

In sentences written in the active voice, action is expressed directly; the subject is doing the acting. *John opened the valve* is active because the subject, *John,* is opening the valve. By comparison, *The valve was opened by John* is a passive sentence in that the action is indirect.

Whenever possible, use the active voice. Your writing style will be more direct and vigorous; your sentences, more concise. The passive voice sounds stiff and weak, as you can see in these examples:

Passive	*Active*
Dolphins were taught by researchers in Hawaii to learn new behavior.	Researchers in Hawaii taught dolphins to learn new behavior.

Control of the furnace is provided by a thermostat.	A thermostat controls the furnace.
Fuel cost savings were realized through the installation of thermal insulation.	The installation of thermal insulation cut fuel costs.
The instruction manuals are frequently updated by our technical editors.	Our technical editors frequently update the instruction manuals.

The passive voice *is* used, however, when the doer of the action is unknown or unimportant (or less important than the action itself):

The first smallpox vaccination was given in 1796.

It was found that a virus induces cells to make interferon.

Rule 41. Use plain rather than elegant or complex language.

In the foreword to his book *The Solid Gold Copy Editor*, Carl Riblet, Jr., has this to say about simplicity in writing:

·The copy editor believes that an interesting story should be remarkable for its simplicity. For example—he knows that rain falls in drops, and he is dedicated to the idea that to describe rain otherwise, as "precipitation, water droplets condensed from atmospheric water vapor and sufficiently massive to fall to the earth's surface," is unnecessary, undesirable and unendurable.*

Like the newspaper copy editor, the technical writer strives for simplicity by avoiding fancy phrases, bombast, and "purple prose." Plain language communicates more effectively than complex language, and communication—not literary style—is the mark of good technical writing:

Complex	*Simple*
I feel that the application of these key principles will provide me with a clear-cut method of handling problem situations, while affording the employee the opportunity, the experience, and the feeling of interacting with me.	The application of these key principles will give me a method of solving problems and allow me to work closely with the employee.
Another very important consequence of Einstein's theory of special relativity that does not follow from classical mechanics is the prediction that even when a body having mass is at rest, and hence has no kinetic energy, there still remains a fixed and constant quantity of energy within this body.	According to the theory of special relativity, even a body at rest contains energy.
The corporation deemed it necessary to terminate Joseph Smith.	Joseph Smith was fired.

Rule 42. Delete words, sentences, and phrases that do not add to your meaning.

Unnecessary words waste space and the reader's time, and they make strong writing weak. The fewer words you use, the better. After you have written a first draft, go through it with a pencil and strike out all words, sentences, phrases, and pages that do not add to your meaning. Here are a few examples of how concise writing is livelier and more readable than wordy prose:

Wordy	*Concise*
It is most useful to keep in mind that the term *diabetes mellitus* refers to a whole spectrum of disorders.	*Diabetes mellitus* refers to a whole spectrum of disorders.

Anthropologists have long observed that the Jale people, who live and dwell in New Guinea, will exhibit cannibalism in that they eat the bodies of enemies they slay in the conflict of war.	The Jale people of New Guinea eat the bodies of foes slain in war.
In the majority of cases, the data provided by direct examination of fresh material under the lens of the microscope are insufficient for the proper identification of bacteria.	Often, bacteria cannot be identified under the microscope.

As you can see, concise writing is crisper, simpler, and easier to read.

Sometimes deleting whole paragraphs can improve a piece of writing. Here is the lead paragraph from an article published in a leading trade journal:

It is both exciting and rewarding to discover that the scientific principles of one's profession can have immediate and gratifying expression in daily life. A case in point occurred recently, and I think it is appropriate to relate.

Beginning technical writers feel the need to ease the reader into their writing with this kind of lengthy "warm-up" introduction. But this paragraph does not contain news, or important facts, or items of interest. *Of course* the author thinks "it is appropriate to relate"; otherwise, he or she would not have written the article. How much better to delete the unnecessary paragraph and plunge right into the story!

Rule 43. Use specific and concrete terms rather than vague generalities.

When engineers read a brochure or report, they seek detailed technical information—facts, figures, data, recommendations, and conclusions. By its very nature, technical writing must deal in specifics, not generalities.

Below are two versions of a technical advertisement. The one on the left was written by an advertising-agency copywriter; it is clear and concise but a little short on detail. A technical writer's rewrite appears at the right. This advertisement, though slightly longer, is much more persuasive because it deals in hard numbers—dollar savings, on-line availability, and efficiency.

How Our New Scrubber Means Big Savings for Samson Power

Aircom's new scrubbing system saves the Samson Company a fortune in fuel every day.

How?

By allowing Samson to burn cheap, high-sulfur coal instead of expensive compliance oil, the Aircom Scrubber cuts fuel costs.

This new system has proved itself efficient. And reliable. To find out more, send for our free brochure.

Aircom's Scrubbing System Saves Samson $7,000 Every Day

With Aircom's new scrubbing system, Samson can burn high-sulfur coal instead of expensive compliance oil and *still* meet all federal and local emission-control regulations.

The result is a fuel cost savings of $7,000 a day—a 35 percent reduction in Samson's annual fuel bill.

For over one year, the Aircom system at Samson has demonstrated an on-line availability of 98 percent. And an average SO2 removal efficiency of 92 percent.

To find out how this commercially proven scrubber can clean the air and cut your fuel costs, send for our free brochure today.

Do not be content to say something is good, bad, fast, or slow when you can say *how* good, *how* bad, *how* fast, or *how* slow. Be quantitative, not qualitative, whenever possible:

He ran fast.	He ran the 100-yard dash in 10.2 seconds.
The sun is hot.	The sun is hot—almost 11,000°F at its surface.

The words you choose should be specific and concrete:

Our measurements are not precise because the experimental apparatus was in poor condition.	Our weight measurements are not precise because the scale was not working properly.
The expedition was delayed for a time because of unfavorable weather conditions.	The expedition was delayed one week because of snowstorms.

Rule 44. Use terms your reader can picture.

In *Heavy Equipment*, written, designed, and illustrated by Jan Adkins, the author describes the subject in these words:

> They are the big machines, the heavy equipment that chews at the earth and builds on it. They are so strong! The power of a thousand horses lives in their metal hearts, and nothing can stand against them.*

Colorful language, when used properly and sparingly, can create in the mind a picture far more memorable than any catalog of facts and statistics. Technical writers should welcome the picturesque phrase when it helps get the message across. Here's how Russell F. Doolittle, author of the article "Fibrinogen and Fibrin," helps form the concept of platelets in the reader's imagination:

> Prick us and we bleed, but the bleeding stops; the blood clots. The sticky cell fragments called platelets clump at the site of the puncture, partially sealing the leak.†

---

*Jan Adkins, *Heavy Equipment* (N.Y.: Scribners, 1980), p. 38.
†Russell F. Doolittle, "Fibrinogen and Fibrin," *Scientific American*, December 1981, p. 126.

The author could have written, "Platelets help our blood to clot," but by using vivid, concrete terms, such as *sticky cell fragments, clump, site of the puncture,* and *sealing the leak,* Doolittle has created a crisp, clear image of blood clotting, an image immeasurably preferable to a dry recitation of scientific technicalities.

Rule 45. Use the past tense to describe your experimental work and results.

Research reports are written in the past tense because they describe work completed in the past:

> Sensitivity after drug withdrawal began an average of a few days after the last dose and lasted an average of six days.

> The Department of Sanitation measured the flow rate for each of the three pipe lines.

> Samantha used a mixing valve to create a liquid-liquid dispersion.

No one expects you to put laboratory experiments that occurred in the past into the livelier present tense; it would be inappropriate.

Rule 46. In most other writing, use the present tense.

Hypotheses, principles, theories, facts, and other general truths are expressed in the present tense. Avoid using the conditional *could* or *would* and invoking the future tense needlessly, because these uses add an unnecessary sense of indefiniteness to a definite statement:

The changing about of one amino acid in the chain could make an entirely different protein.	The changing about of one amino acid in the chain makes an entirely different protein.
Crystals would form from fusion if the temperature or pressure was high.	Crystals form from fusion at a high temperature or high pressure.

The rocket will usually burn eight times its weight in liquid oxygen.	The rocket usually burns eight times its weight in liquid oxygen.

Rule 47. Make the technical depth of your writing compatible with the background of your reader.

If a technical term makes your writing clearer or more concise, use it:

The chemist poured water into an open glass cylinder with a pouring lip.	The chemist poured water into a beaker.
internal software programs that run the computer system	operating system
small-scale test version of a larger industrial process	pilot plant

But avoid technical terms when a simpler word will do just as well, or when a term's meaning may be unclear to a significant portion of your readers:

The moon was in syzygy.	The moon was aligned with the sun and the earth.
Maximize the decibel level.	Turn up the sound.
Stabilize mobile dentition.	Keep loose teeth in place.

Systems professionals strive to keep their user's manuals and other documentation at the user's level. Just as a parent must learn to explain things to a child on a basic level, so too systems people are faced with having to gear their manuals to people who are unfamiliar with computers and perhaps even computer-phobic. The inexperienced technical writer may think, "The user should know how to do it!" but the best writers meet the readers' needs—even if the reader is at a basic level.

Rule 48. Break up your writing into short sections.

In technical writing, it is not uncommon to find sentences of more than 40 words, or paragraphs that run on for pages. While some ideas may truly need this type of elaboration, a great many do not.

Readers like brevity, so you should look for ways to break long sentences and paragraphs into shorter, easier-to-grasp units.

In any 50- or 60-word sentence, there's at least one opportunity to divide the sentence into two shorter sentences. And a paragraph of 10 to 15 sentences might be broken into two or more shorter paragraphs by finding places where a new thought or idea is introduced and beginning the new paragraph with that thought.

In the same way, the text itself should be broken up into many short sections and subsections. Use headlines and subheads to title each section. Each short section should be a self-contained mini-essay on a single idea or thought. Experience teaches us that a piece of writing is most effective when each section deals with one topic.

Rule 49. Keep ideas and sentence structure parallel.

Parallel sentence structure exists when two or more sentence elements of equal importance are similarly expressed. The benefits of parallelism are many: an economy of words, a clarification of meaning, a sense of symmetry, and a sense of the equality of each idea in the sentence. In fact, the previous sentence is an example of parallelism.

In the sentence *The tube runs into the chest cavity, across the lungs, and into the stomach,* it is the similar construction of the three prepositional phrases that makes for parallel construction. Too often, writers violate parallelism by constructing part of the sentence in one way and switching to another construction later in the sentence, as in the example on the left:

Nonparallel	*Parallel*
Please sign the proposal, date it, and it must be sent to me.	Please sign the proposal, date it, and send it to me.

Here are a few other examples of parallel construction:

The atomic weight of gold is 196.97; silver, 107.87; iron, 55.85; lead, 207.19.

Ask not what your country can do for you; ask what you can do for your country.

It was the best of times; it was the worst of times.

Rule 50. Opt for an informal rather than a formal style.

Do not hide behind an overly formal writing style. We are not recommending that you write your technical papers in the same chummy tone as you would a letter to your pen pal. But a more relaxed, conversational style *can* add conviction, readability, and vigor to your work.

Technical writers go to great lengths to avoid using personal pronouns. This can result in an unnaturally stiff style. Why conceal your identity? Instead of writing *the measurement was taken,* write *I took the measurement*—if, indeed, you did take the measurement. Here are some examples of formal versus informal style:

Formal	*Informal*
It is unfortunate that I was not available when you visited our facilities the other day.	I'm sorry I missed you the other day.
For the purpose of breaking up a beam of sunlight into the seven visible colors of the spectrum, a glass prism was procured.	I used a prism to break up sunlight into a rainbow.

In particular, avoid dated transition words such as *whereby, heretofore, herein,* and *wherein.* If your writing reads as if you went to law school, you have fallen under the spell of "legalese" (also known as "bureaucratese," "federalese," or "corporitis"), in which, for no apparent reason, words take on a certain institutional stiffness. When words such as *whereby, thereby, heretofore,* and *wherein* creep into your vocabulary, put down your pen, take a few deep breaths, and read your work aloud. Your ear will soon tell you just how awkward and antiquated these phrases are:

Perchance	This sounds like something Shakespeare might have said almost 400 years ago ("to sleep, perchance to dream"); it is a bit poetic for technical writing. Write *maybe* instead.
Hitherto	Why not simply use *until now?*
Enclosed herewith	Here*with? Where*with? It sounds pretentious as well as starchy. Use *enclosed* alone.
Inasmuch as	Archaic. Use *since* or *because.*
Attached hereto	Stilted, awkward.
Enclosed herein	Stilted.
Whereof	Stilted.
Thereof	Stilted.
Thereby	Stilted.
Thereto	Thereto? *Where*to?
Of even date	Insurancese for *today.* Why not just write *today?*
Pursuant to your orders	Overly formal. Just write *following your directions* or *as you instructed.*

Prior to	Use *before*.
Keep me timely advised	Insurancese for *let me know as soon as you do*—a phrase that at least tells laypeople what is going on!
Aforementioned	Unless you write leases, avoid this lawyerlike expression.
Etc.	*Etc.* is fine when used in a sentence in which the reader can predict its meaning: "Let's discuss even numbers, like 2, 4, 6, etc." Often, though, it is a lazy way out of a thought (e.g., "Go to the laboratory and get me a bunsen burner, flask, etc.").

5

Words and Phrases Commonly Misused in Technical Writing

Technical Words and Jargon

Every field has its own technical vocabulary, a language that helps specialists describe a concept, process, or thing. The term *standard deviation* has a precise meaning to statisticians, in the same way that *excess* has specific meaning to an insurance claims adjuster.

Technical words such as these are helpful and necesary. Problems arise, however, when technical words proliferate: Some words become a slanglike shorthand for precise communication. At that point, technical terms turn into buzz words, and these catchphrases hoodwink writers, making them forget that the words have limited application and may be understood by only a few insiders.

Don't throw around jargon. A hospital administrator might use the term *catchment area* to describe the place from which hospital patients are drawn. But this term sounds strange to people who are not in the health services and may cause an outsider to be confused (and possibly to giggle).

In the same way, how much is the technical writer communicating to the nontechnical reader when, in an automobile advertisement running in *Time* magazine, she or he writes that a particular model has an "electronic instrument cluster, an electronic bar chart fuel gauge, an optimum 3.8 liter V-6 engine, an automatic overdrive transmission, a modified MacPherson strut front, supervision, and stabilizer bars, and 4-bar link rear suspension"?

Even everyday words can have many meanings to people in different specialties. To doctors, a *tongue* is what they tell patients to stick out of their mouths. But concert organists know tongues as vibrating slips of metal producing the tones in their instruments. Likewise, if you say *drone*, biologists think of a male honeybee, military personnel of remote-control aircraft, and Scotsmen of the continuously sounding pipes in their bagpipes.

Technical writers need to decide when they are using appropriate terms and when they are obscuring their meaning in needless slang or bombarding the reader with technical overkill.

Do not invent technical words just to add a touch of importance to your writing. We once saw a memorandum dealing with park "signage," and it took us several minutes to figure out that the subject was signs. Inventing and perpetuating terms such as *on-line, outage, linkage, stagflation, gridlock, feedback,* and *cost-effective* may be fun—especially for the media—but the careless writer may forget that these expressions are not universally accepted and understood. If two computers are not interfacing, it may be a serious problem; but when a writer states that two executives don't interface, we worry more about the creator of the phrase than about those poor noninterfacing executives!

When words or phrases such as *meltdown, half-life,* and *burnout* become ingrained in a writer's vocabulary, they begin the long, slow slide into clichédom and oblivion. While they are on their way, these words tend to alienate readers from the precise meanings they only hint at.

Once you get over the temptation to make your writing sound important and to be a verbal show-off, you will have taken a big first step toward simplicity and clarity. A media advertisement that we saw recently serves as a good example of unnecessarily technical language. It described a drug that shows a "low incidence of adverse reactions." Translated, this means that the drug has few side effects.

Big Words

Technical writers sometimes prefer big, important-sounding words to shorter, plainer words. As the number of syllables grows, the writing may sound more "important," but it becomes harder to understand. When sentences get long and words balloon in size, the mind of even the most patient and intelligent reader begins to wander.

If technical writers could achieve some aesthetic distance from their writing and put themselves in the reader's place, they would search for simple words instead of long, complicated ones to express their thoughts. Too often, technical writers forget that their audience may include people outside the writer's department, industry, or discipline.

In electronics, for example, technicians do not *free* a link; they *disengage* it. Laboratory experiments are never *ended*, they are *terminated*. And, of course, engineers never *estimate*, they *approximate*. (Even airline personnel prefer the cumbersome term *deplane* to *leave* or *get off*.)

Do not use a big word when a smaller one will do.

Here are a few big words in common use. The column on the right shows a simpler—and preferable—substitution:

Instead of:	*Use:*
abbreviate	shorten
aggregate	total, whole
amorphous	shapeless
anomalous	abnormal
antithesis	opposite

aqueous	watery
ascertain	find out
autonomous	independent
beverage	drink
cessation	stop, pause
circuitous	roundabout
coagulation	clotting, thickening
comestibles	food
commencement	start, beginning
concept	idea
conjecture	guess
contiguous	near, touching
currently	now
deficit	shortage
demonstrate	show
discourse	talk
disengage	free
duplicate	copy
eliminate	cut out
elucidate	clarify
expedite	hasten, speed
facilitate	ease, simplify, help
feasibile	possible
gradient	slope
homogeneous	uniform, similar
impairment	injury, harm

incision	cut
incombustible	fireproof
inundate	flood
maintenance	upkeep
minuscule	tiny
nomenclature	name, system of terms
obtain	get
optimum	best
orientate	orient
parameter	variable, factor
posterior	rear
potentiality	potential
requisite	needed, necessary
segregate	set apart
subsequent	next
sufficient	enough
terminate	end
verification	proof
viable	workable
vitreous	glassy

Wordy Phrases

Avoid the wordy phrase; strive to be succinct. During a 1981 World Series game, sportscaster Howard Cosell commented that the Yankee owner and the manager had "a mutuality of affection for each other." We suppose that means they liked each other.

When you edit your writing, simplify those wordy phrases that take up space but add little to meaning or clarity. The following list includes some common wordy phrases. The column on the right offers suggested substitutions:

Instead of:	*Use:*
a large number of	many
along the lines	like
as a general rule	generally
as shown in table 6	table 6 shows
as yet	yet
at all times	always
at this point in time	at this time, now
at your earliest convenience	now, soon
be considered as	is
by means of	by
despite the fact that	although, even though
during the course of	during
even more significant	more significant
exhibits the ability	can
has been widely acknowledged as	is
has proved itself to be	has proved, is
have discussion of	discuss
hold a meeting	meet
inasmuch as	since
in many cases	often
in order to	to
in some cases, in other cases	sometimes

in the course of	during, while
in the event that	if
in the form of	as
in the majority of instances	usually, generally
in the near future	soon (state approximate or exact date)
in the process of tabulating	in tabulating
in the vicinity of	near
is equipped with	has, contains
it is clear that	clearly
on a daily basis	daily, every day
on a weekly basis	weekly
on an annual basis	yearly
on the basis of	by, from
on the occasion of	when
prior to that time	before
start off	start
subsequent to	after
take action	act
the necessity is eliminated	you do not need to
the reason why is that	because
until such time as	until
with reference to	about
with the result that	so that

Some phrases are so inflated that they can be omitted entirely. They merely take up space and should be deleted from every sentence:

deemed it necessary to
it has been shown that
it is found that
it is recognized that
it is the intention of this writer to
it is worthy of note that
it may be mentioned that
it may be seen that
it must be remembered that
it will be appreciated that
thanking you in advance for your cooperation
the fact that
what is known as

Redundancies

Redundancy is an insidious form of wordiness that inflicts itself on the work of nontechnical as well as technical writers. Some redundant words are modifiers that merely repeat an idea already contained in the word being modified.

Notice how many people use the phrase *very unique*. Actually, *unique* means one of a kind, so it is impossible for anything to be *very* unique.

One car manufacturer designed its advertising campaign around the slogan *new innovations*. Is there such a thing as an *old* innovation?

The best way to spot redundancy is to ask what a word is "buying" in a particular phrase. Is it *adding* meaning? Or is the word there because the writer was not sure that the thought had been communicated completely by using *unique* or *innovation* or any other sharply defined terms?

Consider the phrase *study in depth*. Ask yourself: Doesn't the word *study* already imply *in depth?* In the same way, *consensus of opinion* is a redundancy because *consensus,* by itself, implies a solidarity of group opinion.

When redundancies are examined in this way, writers can spot them easily. A phrase such as *investigative reporter* suddenly sounds odd, because all reporters investigate. Could there be such a thing as a *non*investigative reporter?

Here's a list of other common redundancies, along with suggested substitutions:

Redundancy	*Substitutions*
absolutely essential	essential
absolutely perfect	perfect
actual experience	experience
adding together	adding
advance plan	plan
all of	all
an honor and a privilege	an honor
any and all	any
balance against one another	balance
basic essentials	essentials
by means of	by
cancel out	cancel
combine into one	combine
consecutive in a row	consecutive
continue on	continue
cubic meters in volume	cubic meters
current status	status
different varieties	varieties
equally as well	equally
final outcome	outcome
first and foremost	first
first introduction	introduction

first priority	priority
goals and objectives	goals
Gobi Desert	Gobi
honest truth	truth
in close proximity	close
isolated by himself	isolated
joined together	joined
main essentials	essentials
mixed together	mixed
mutual cooperation	cooperation
necessary requisite	requisite
one and the same	the same
open up	open
overall plan	plan
past history	history
personal opinion	opinion
physical size	size
point in time	time
rain shower	shower, rain
reason why	reason
refer back to	refer to
repeat again	repeat
small in size	small
take action	act
this particular instance	this instance
this particular time	now

triangular in shape	triangular
true facts	facts
uniformly consistent	consistent
whether or not	whether
wrote away for	wrote for
you may or may not know	you may know

Clichés

Clichés are words or phrases that have worn out their welcome—in fact, *worn out their welcome* is a good example of a cliché! They have become trite through overuse, and so they no longer communicate in the same fresh way they once did. In a sense, they have been penalized for their popularity. Here are a few clichés that should be used sparingly if at all:

acid test
back to square one
ballpark
beyond the shadow of a doubt
bottom line
bread-and-butter issue
cost-effective
dealing with
eyeball (used as a verb)
feedback
great success
grind to a halt
hands-on
light of day
meaningful
no-no
overkill
point in time
richly deserved
state of the art

top dollar
tried and true
try it on for size
under review
viable
vitally important

Overblown Phrases

Words reflect society, and when society changes so do its words. Some phrases, however, seem to linger. Perhaps these throwbacks to an earlier time have become so comfortable that we perpetuate them instead of retiring them.

You can spot an overblown phrase—whether it's antiquated, pompous, or just a meaningless stock phrase—by reading your writing aloud. When we read aloud the following phrases, we are reminded just how ludicrous, old-fashioned, or pompous they have become:

I'm sure you can appreciate	This phrase is patronizing and should be avoided
note how this matter will be handled	Also patronizing.
when time permits	It may sound poetic, but it's also inaccurate. Time doesn't permit; people do.
by virtue of	To paraphrase the late Mae West, "virtue" has nothing to do with it. *Because* is usually an effective substitute.
kindly advise	As opposed to *unkindly?* Unnecessary.
don't hesitate to	Unnecessary and wordy. Instead of writing, "Please don't hesitate to call me," write, "Please call me."

deemed it necessary   Old-fashioned.

under separate cover   Also old-fashioned. Use *sep-
                               arately.*

The Rise of *-Ize*

The tendency to add *-ize* to nouns is an old story in English. The
practice has been going on for centuries. *Apologize* was born
before 1600, and *criticize* appeared in Shakespeare's day. *Rev-
olutionize* came along before 1800; *burglarize* first appeared in
the 1870s. Sixty years ago, people spoke of *Broadwayizing* a
play—but the term, fortunately, came and went. A mailing pro-
moting a business seminar introduced *four* new *-ize* words: *cus-
tomerize, projectize, expertize,* and *marketize.*

Edwin Newman, in his book *A Civil Tongue,* suggests that by
adding *-ize* to certain words, people believe they are achieving
a more businesslike (i.e., "professional") tone. *Prioritize* may
sound more businesslike than *make priorities,* but it is also awkward, pretentious, and incorrect. *Utilize* may sound more technical than *use,* but most of the time the latter is the better choice.
The head of Turner Broadcasting may use the word *colorize* to
describe changing black-and-white films into color, but the word
he really wants is *color.*

Some *-ize* words are ambiguous. Take *finalize.* Does this mean
to complete? If you finalize my contract are you signing it? Or
are you just about to sign it? By using *finalize,* writers slip into
vagueness by failing to tell us precisely what has been concluded.
It could be the signing of a document, the agreement to the
wording, the agreement to even write a contract. Therefore, the
finalizing of a contract is too vague and general to pass along
clear meaning.

Here are a few other *-ize* words to be used sparingly, if at all:

academize
customize
formalize
maximize

normalize
optimize
politicize
standardize
traumatize
utilize

Nouns as Adjectives

Strictly speaking, it is poor writing to use a noun as an adjective. But today, it's common practice; using nouns as adjectives can make a sentence more direct, less wordy. For example, the phrase *data analysis* uses the noun *data* as an adjective modifying *analysis*. It's better than the more roundabout phrase *analysis of data*.

Nouns as adjectives are acceptable when used sparingly and clearly. In technical writing, the trouble starts when writers string nouns together to make cumbersome, hard-to-follow phrases.

We have all had the experience of watching adjectives proliferate until they actually obscure, instead of illuminate, meaning. For example, if you were describing a power boiler at an industrial plant, you might write "an industrial power boiler." Let's say you had to include the power rating and fuel. Now you write "a 15,000-lb steam/h pulverized-coal-fired industrial power boiler." If too many adjectives pile up, break up the list with prepositional or participial phrases:

a 15,000-lb steam/h pulverized-coal-fired power boiler	a pulverized-coal-fired power boiler generating 15,000 pounds of steam per hour
chronic post-traumatic stress disorder	chronic stress disorder following trauma
a frequency-shift power-line carrier relaying system	a relaying system using a frequency-shift power-line carrier

| software-programmable modular information-retrieval system | a modular informative-retrieval system with programmable software |

Misused and Troublesome Words and Phrases

English is a rich language, and so it is inevitable that words with subtle shades of meaning are often confused and misused. Some words are misused because slang has made their meanings muddy; others, because their actual definitions are overshadowed by what the public believes the words mean; still others, because they sound alike and are often mistaken for one another.

The following is a selected list of commonly misused words and phrases. Since the technical writer generally must uphold a higher standard of accuracy than the nontechnical writer, you may wish to review these words by studying their definitions and using them in sentences:

ability, capacity	*Ability* means the state of being able, or the power to do something. (A computer has the ability to create graphics.) *Capacity* is the power of receiving or containing. (The computer has a capacity of 640 K.)
about, approximately	*About* indicates a rough estimate. (We are about halfway there.) *Approximately* implies near accuracy. (There are approximately 1.05 quarts in a liter.)
accept, except	*Accept* means to receive willingly, to agree with. (I accept your apology.) *Except* means excluding. (You'll be reimbursed for everything except local travel.)

advise, inform	*Advise* means to offer counsel and suggestions. (I advise you buy a mutual fund.) *Inform* means to communicate information. (I inform you that your proposal hasn't arrived yet.)
affect, effect	*Affect* is a verb meaning to change or influence. *Effect* is a verb meaning to bring about. *Effect* is also a noun meaning result or outcome. (The report will have the desired effect.)
aggravate	*Aggravate* means to make worse. Don't use it as a synonym for irritate, annoy, or provoke. (The layoffs will only aggravate the problem.)
alternate, alternative	*Alternate*, as a noun, means a substitute. An *alternative* is a choice between two or more possibilities.
and/or	This is an awkward construction. Avoid it.
anxious, eager	Use *anxious* when anxiety or worry is involved, not as a synonym for eager. (I'm anxious about my performance appraisal.) *Eager* means highly desirous of something. (I'm eager to know the results of our work.)
because of, due to	*Because of* means by reason of, or on account of. (The conference was delayed because of

	snow.) *Due to* means attributable to. (Her promotion was due to her managerial style.)
beside, besides	*Beside* means by the side of. *Besides* means in addition to.
between, among	Use *between* when writing of two things. Use *among* when writing of three or more things.
big, large, great	*Big* is used in connection with bulk, mass, weight, or volume. *Large* is used to describe dimensions, extent, quantity, or capacity. *Great* is now used almost entirely to connote importance, eminence, superiority, or excellence.
can, may	*Can* implies ability; *may* implies permission or possibility.
capital, capitol	A *capital* is a town or city that is the official seat of government in a political entity. *Capital* can also refer to an upper-case letter or to money (e.g., capital needed to start a business). A *capitol* is a building in which a legislature assembles.
center, middle	These terms are not interchangeable. From its geometric definition, *center* retains, even in nontechnical contexts, the idea of a point around which everything revolves or

	rotates. *Middle* is less precise, suggesting a space rather than a point.
continual, continuous	*Continual* means recurring frequently; *continuous* means without interruption.
convince, persuade	*Convince* means to overcome with proof. *Persuade* is to plead or urge through argument.
data, datum	When *data* is used synonymously with facts, it is plural. When it is used synonymously with information, it is singular. The singular form *datum* has fallen out of popular use.
disinterested, uninterested	*Disinterested* means impartial. *Uninterested* means indifferent, bored.
effective, efficient	A machine that's *effective* performs its intended function well. If it does this with a minimum of waste, expense, and unnecessary effort, then it's *efficient* as well.
e.g., i.e.	*e.g.* means for example; *i.e.* means in other words, or that is.
ensure, insure, assure	*Ensure* means to make sure of something. *Insure* means to take out an insurance policy. *Assure* means to inform confidently.
equipment, equipments	Use *equipment* as singular or plural. There is no such word as *equipments*.

farther, further	*Farther* refers to physical distance. (He is farther away from the plant than he is from headquarters.) *Further* refers to matters in which physical measurement is impossible. (Further research would be helpful.)
fewer, less	*Fewer* is used when units or individuals can be counted (fewer memos). *Less* is used with quantities of mass, bulk, or volume (less space).
finalize	Use *complete* or a more specific term. (Not: We are going to finalize your contract. Instead: We are going to sign your contract.)
hopefully	*Hopefully* means in a hopeful manner. The sentence *Hopefully the situation will improve* makes no sense because the situation cannot be full of hope. *Hopefully we shall fly to Pittsburgh tomorrow* does not mean we hope to fly to Pittsburgh tomorrow. It means we shall fly there full of hope. Beware *hopefully*.
impact	Do not use *impact* as a verb to mean to affect or influence. To *impact* means to drive or press closely into something. If you want to say X had an impact on Y, say X affected, influenced, or impinged on Y.
irregardless	Not a word. Use *regardless*.

like, as	*Like* means similar to. It is still not acceptable as a conjunction. It is acceptable, however, when it introduces a noun not followed by a verb. (This coffee is like an espresso.) *As* means in the same way that. (Think as I think.)
matériel, material	*Matériel* is the equipment, apparatus, and supplies used by an organization. *Material* refers to the substances of which something is composed.
over, more than	*Over* implies position. Do not write *over* when you mean more than. (There are more than 200 branches nationwide.)
percent, percentage	*Percent* means per hundred. *Percentage* means a proportion or share in relation to a whole.
practicable, practical	*Practicable* means capable of being put into practice. *Practical* describes something that can be done based on past performance.
presently, at present	*Presently* means soon; *at present* means now.
principal, principle	*Principal* used as a noun means head of a school, a main participant, or a sum of money. As an adjective, it means first or highest in rank,

	worth, or importance. A *principle* is a fundamental law, a basic truth.
prioritize	Not a word. Use *make priorities* or *order*.
should, will	*Should* implies ought to, or suggests a belief. *Will* suggests intention.
strategize	Awkward. Use *make strategies*.
that, which	Ideally, *that* is used with a restrictive clause—a clause absolutely necessary to the sentence. (This is the project that will launch your career.) *Which* is used with a nonrestrictive clause—a clause that adds descriptive matter but is not necessary to the sentence. (The executive committee, which is made up of vice presidents, has not discussed the problem.)
ultimate, penultimate	*Ultimate* means last. *Penultimate* means next to last. Do not use penultimate as a superlative of ultimate.
unique	No superlatives are needed, since *unique* means one of a kind. Therefore, really unique, and similar modifications are grammatically incorrect.
utilize	Use *use*.
who, whom	Use *who* as a substitute for he, she, or they. (Who will be the

boss—Bill or Sheila?) Use *whom* as a substitute for him, her, or them. (To whom shall I bill the room charge—him or her?)

In Conclusion . . .

In this chapter, we've covered the use of fewer than 200 words. Your vocabulary may be as large as 20,000 words or more, so not all your questions about proper word choice will be answered by the examples we've chosen.

Therefore, when you question the choice of a particular word in your writing, ask yourself three questions:

- Is there a shorter, simpler, or more modern word that would get my meaning across just as well?
- Will most of my readers understand the word?
- Is the word as specific and concrete as it can be?

If the word is too complex, too long, unfamiliar, old-fashioned, vague, or abstract, search for another word that will serve you better.

PART II

Tasks of the Technical Writer

6

Proposals and Specifications

Proposals

What Is a Proposal?

Simply put, a proposal is a plan offered for acceptance or rejection. An engineer might propose the replacement of a boiler in a pulp and paper mill. A systems analyst might propose a piece of new software for his or her department. An aerospace company might assemble a team of people to write a proposal to build a new billion-dollar fighter-bomber for the US Department of Defense.

Types of Proposals

Proposals come in different formats and sizes to fit different situations.

Letter proposals are informal, can be brief (usually two to five pages), and contain information about the project plan, staffing, and budget. They may also include a brief statement that requests a positive step—for example, asking the reader to sign the letter as authorization to begin work.

Feasibility reports are similar to proposals except that they evaluate two or more possible solutions to a problem, where a proposal usually recommends one solution—the one the proposal writer is trying to win approval for. The most common organization for a feasibility report is:

- table of contents
- list of figures
- abstract
- introduction, in which you state the purpose of the report, describe the problem the report addresses, and present the scope of the report
- background
- criteria
- possible solutions
- conclusions
- recommendations

Formal proposals, often written in response to a Request for Proposal (RFP), will generally have the following organization:

- title page
- table of contents
- executive summary
- introduction
- background
- discussion
- project organization with timetable
- budget
- qualifications and experience
- summary
- appendix

Since there are whole books devoted to the craft of proposal writing, we will not attempt to summarize the intricate and diverse ways in which informal and formal proposals are written. We will, however, set forth a few principles of proposal writing, tips on analyzing RFPs, ways to write more sales-oriented headlines and subheads, and observations concerning those who read proposals. We'll conclude with commentary about how proposals are evaluated.

Principles of Proposal Writing

Everyone who has written a proposal has probably formulated some do's and don'ts about the process. We've come up with 10 principles that we feel can be applied widely:

1. *Learn everything you can about your prospective client or the people who will evaluate your proposal.* Every scrap of information about the prospect can be of help. What are the goals of the project they propose? How have similar projects at their organization succeeded? Are they strapped for cash, or is money not a key factor in determining who will be awarded the project?

Often there is a person who represents the prospective buyer (sometimes referred to as the "point of contact") who will respond to questions from those submitting proposals. Call that person; you'll be amazed at the details you can learn just by asking a few questions.

Once, we called a point of contact and said, "It sounds as if you're looking for a consultant to spend a month in Florida writing a manual." "Oh, no!" said the point of contact. "We think the job should be completed within two weeks." By learning of the unstated boundary (two weeks), we were able to reformulate our proposal accordingly. But we would never have learned about the two-week limit without making the call.

2. *Sell your ideas by fitting them into the client's needs.* There is often a big difference between what you have to offer and what the prospective client needs. Technical people frequently fall into the habit of focusing all their energy on solving the technical problems in the project. But we believe that you must first show empathy for the client's needs, then fit your ideas into that framework. Every time you start going into the details of a solution to a problem, put yourself in the client's shoes, asking, "How does this solution help him (or her)?" Keep reminding the

client how your work will, as its ultimate goal, solve his or her problem. For example, instead of:

We have extensive experience in airlines operations and forecasting and evaluating traffic flows.

Write:

Our extensive experience in airlines operations and forecasting and evaluating traffic flows gives us insight into the logistics of your business and will help you respond faster.

3. *Don't just solve the technical problem; empathize with the customer's critical needs.* This is slightly different from the previous tip. Customers may be so concerned about cost overruns that this concern outweighs their interest in the precise method you choose to solve their problem. You must reflect the client's critical needs: In other words, by rephrasing their needs in your own language, you are telling clients that you are listening and that you care. This empathizing is a crucial first step before sailing into the actual solution suggested in your proposal.

4. *Recognize all critical factors that evaluators use in assessing proposals.* When evaluating vendors, an organization usually has its own criteria for deciding which to use. Price may be one critical factor, but there may be many others. Some companies like to see a long track record of reliability. Others want to know of your competence in handling similar projects in the past. Still other companies will place the highest value on getting the project completed quickly.

We once asked a point of contact if he would give us his own criteria for judging proposals. Surprisingly, he mailed us a sheet containing a series of questions that guided him through his reading of each proposal. One question he expected to see answered in many of the proposals he read was, "What would be the downside of doing nothing to solve the problem?" In other words, he expected proposal writers to look into all alternatives—including the client's option of not completing the proposed project.

5. *Make sure your proposal addresses every element mentioned in the RFP.* Local government, for example, can be quite meticu-

lous in its requirements. Something that may seem petty or small to you can loom large to a bureaucrat checking your proposal. Make sure you address all elements of an RFP—not just the most important or obvious ones. Some veteran proposal writers place an RFP checklist at the front of their government proposals to let the reader know precisely where in the proposal each RFP element has been addressed.

6. *Use appropriate graphics to highlight your ideas and make them easy to visualize.* Paragraphs are wonderful for discussions of ideas, but when those ideas would be unwieldy if expressed in prose, look for alternatives: charts, graphs, illustrations, bulleted lists, numbered items, and tables. But just because you choose a graphic, don't assume that it will automatically be clear to the reader. Keep charts, graphs, and tables simple. If you put too many lines on a graph or too many bullets in a list, you run the risk of overwhelming your reader.

7. *Tailor each proposal to the needs of the specific client.* When you write proposals, you may have on file (or in a word processor's memory) several handy formats and examples of old proposals. While these may help you in gathering your thoughts about formatting the proposal at hand, don't be tempted to re-cycle large amounts of the old verbiage in your new proposal.

Why? Because potential clients like to think of themselves as unique and their needs as unique. They don't want to read a proposal offering "off the shelf" solutions to these unique problems. So, try to weave your newfound information about the current prospective client into the proposal. Everything—even your biography or résumé—may need to change from proposal to proposal; manufacturing clients will want to read about your manufacturing experience, but food-service clients will want to know about your expertise in that area.

8. *Anticipate and defuse objections.* Often, you have insight into how your competitors will propose to solve a potential client's problem. You may have a better way. Your task is to let the proposal reader know that, despite some commendable ideas in your competitors' outlooks, you have the best overall solution. For example, we give on-site writing seminars, and every now

and then a company asks us to propose how we could meet the training needs of several hundred employees. We're aware that companies offering video writing programs will be bidding against us. While we recognize the low cost of video programs compared with on-site seminars, we favor the live instructor.

Therefore, in our proposal, we would bring up the idea of solving the problem via video . . . only to reject the notion because, despite the low cost, video programs never quite get to the heart of what makes a person's writing poor. Also, video programs can't respond to individual problems and questions. For this, we believe only a live (and inspiring) instructor will do.

9. *Avoid hedging and subtlety in proposals.* Proposal readers can see the weakness in phrases like *in most instances, probably, under some circumstances,* and their brethren in hedginess. Proposal readers want to read authoritative-sounding proposals. Being wishy-washy will not be interpreted in your favor.

In the same way, while humor can be delightful, leave it out of proposals. You can never tell if your audience will get the joke or puzzle over it. Therefore, play it straight. Your job is to inform and persuade—not to entertain.

10. *Make a list of where key resources are located if you don't have a proposal library.* Even if you just create a file folder, clear out a desk drawer, or prepare a separate floppy disk, keep information about past proposals as well as information likely to be of use to you in writing future proposals.

You may want to keep previously generated graphics or even the annual reports of potential client companies. You may also wish to keep a client's newsletters, or periodicals in your field. Keep anything that you feel may have bearing on the success of a future proposal so that you can orient yourself quickly to new projects and learn from past mistakes.

Tips on Analyzing RFPs

If you write proposals in response to Requests for Proposals, here is a short list of suggestions you may find helpful:

1. *Read the RFP thoroughly, and more than once.* Since RFPs are usually put together by committee, you may be able to uncover between-the-lines hints that will suggest ways to approach the project. At the first reading of an RFP, jot down your thoughts as they occur to you. You'll be at your freshest on your first reading.

2. *Prepare checklists* that will force you to read with care, ferreting out specific items on your list. Checklists help you focus on exactly what is required and therefore on which items must be addressed in your proposal. If you don't keep track of the requirements of the project, you may forget to address those requirements when you write the proposal. We've seen people lose projects because they forgot to address a small item mentioned in the RFP.

3. *Don't assume that anything not stated is simply not present.* Read between the lines. Sometimes writers of RFPs are simply weak writers who do not state things carefully. Other times, political considerations prevent the client from writing plainly (e.g., a client may be trying to find a hedgy way of stating that a consultant cannot charge for travel time or that one must stay at an inexpensive hotel while working on the project).

4. *Reflect an understanding of the problem (and possible solutions) before offering your proposed solution.* Often, RFPs suggest solutions that you may find inappropriate to the client's problem. If you can identify the hazards of that solution, you can set yourself apart from other proposers.

5. *Be open to new ways of tackling scheduling problems.* Just because the RFP requests a certain schedule or manner of delivery doesn't mean you are forbidden from offering alternatives.

6. *Be prepared not to bid.* Use some consistent process by which you determine whether you even want to bid on an RFP. Some RFPs require more time or effort than they are worth.

Headings

As you prepare to write your proposal, your head may be swimming with ways in which to solve the technical problems posed by the RFP. But, while the details about how you'll solve the problem may be interesting and important, your potential client is even more interested in the benefits to be derived from those solutions.

Using headings that underscore those benefits is a clever way to transform your proposal from a technical document into a selling document. When you use standard headings like "Introduction," "Executive Summary," or "Conclusions," you are doing only half the job: You're labeling what comes next. But you are not putting a sales spin on those thoughts unless you use your headings to tell the readers what benefits are embedded in your ideas. There are several types of headlines that serve this purpose.

The "how to" headline or subhead in your proposal is useful when you wish to offer your reader a specific, practical benefit. For example, take this headline: "How to Create Low-Cost, Highly Customized Software." This headline does more than just say "Introduction"; it makes a promise to the reader. Another how-to headline: "How to Save 13 cents/lb on Palm Kernel Stearine." A final example: "Improve Technical Writing for Pennies a Day." Notice that you do not even have to use the words *how to* in this type of headline.

The question headline uses a rhetorical question to help pull the reader into the headline. For example: "Can XYZ Corporation provide safety and security for your data center for less than $10,000 a year?" The headline forces prospects to read by posing a question that is of interest to them. Another example of this type of headline: "Are you overpaying for your mail-merge software?"

The "reason why" headline helps explain a product or service quickly, so you can distinguish your proposed ideas from those of your competitor. For example, "Five Reasons Why the Department of Defense Should Choose the Hopkins Approach to Satellite Development," or "Three Benefits in Upgrading the CUBE Assembly Line Now."

The command headline energizes and directs readers. "Take the Lead in Underseas Technology" is a good example of a command headline used in a proposal to the US government. "Make the Proposal-Writing Task a Model of Efficiency" is another command headline.

"Direct" headlines promise the reader something new and important. Three recent examples drawn from technical proposals are: "Ongoing Updates to Assure Tailored Upgrading," "An Economical Facility with a Proven Track Record," and "An Efficient Way to Cut Our Manpower Needs by 60 Percent."

How do you come up with selling headlines? A good way is to first write the proposal using ordinary subheads (e.g., "Introduction," "Scope," and so forth). Then, when you edit, study what you've written to see what is the most vital benefit within the section. And that, of course, is the benefit you should highlight in your heading.

How long should headlines be? Don't worry if they're longer than the generic titles. The whole point is that they're generating reader interest. And if your headlines appeal to the personal interest of your potential client, then length is of little importance.

Titles and major headlines should accompany every important point in your proposal. Make sure they accomplish these three things:

1. give an outline of your proposal
2. highlight every important point in your proposal
3. offer good sales arguments for your proposal

Specifications

What Is a Specification?

A specification (or spec) is a description of work to be done. You may write specifications for an air-conditioner, a volleyball court, a lampshade, a milk container, or an airport ground-control system. By writing a precise specification, the technical writer aids in the creation of a product that does what it's designed to do with maximum efficiency.

Although specifications are written by technicians in a variety of fields, we will discuss specification writing as it applies to one large and important technical area: data processing. In the data processing and computing field, specifications often define how a particular process will be automated (e.g., how an ATM will automate certain banking functions; the spec becomes a blueprint for the computer program that will allow the ATM to help banking customers make deposits and withdrawals).

Just as a spec can be written for software to run an automated teller machine, one can also be written to define the workings of a piece of software to handle a company's payroll or to operate a satellite. The main purpose of a software specification is to record the needs and requirements of the intended customer or end user of the system to be developed.

Why Write a Specification?

Sometimes engineers design a system to meet a customer's requirements without writing a specification. That may be appropriate for simple systems or programs that may be used only once. But large, complicated systems contain more information than anyone can remember and involve many end users and developers. The resulting systems may have a long lifetime over which they will be modified and expanded. A specification, therefore, often requires flexibility in allowing for the future growth of the system it is defining.

If upper management at a bank asks the systems department to create an ATM, the resulting specification will be a formal statement of what the managers require the ATM to do. By laying out these requirements, the specification writer helps the managers to refine, change, or add to their original statement of needs. Since this project may be quite costly, the managers need to confront any ambiguities, inconsistencies, or additional customer requirements early in the development of the software. So, a well-written spec can help the bank avoid extensive and costly debugging and modification later on.

How to Develop Specifications

There are four basic steps in the process of writing specs.

The first step is to *get a statement of the problem and requirements from the customer.* In the systems environment, this is called the System Requirements Specification. If you are writing this kind of spec, be sure to state the problem and define the specific system needs. For example, for an automated teller machine, your customer, the XYZ Bank, has this problem statement: "Our bank needs a way to provide our clients with most banking services 24 hours a day. We have decided that an automated teller machine (ATM) will meet those needs, and we need software to run it."

The second step is to *analyze and classify the problem and requirements.* Decide on the minimal set of basic functions, and try to specify them first. Basic functions for the ATM could be deposits, withdrawals, and transfers. From the customer's statement, you find that one requirement is to provide clients with most banking services 24 hours a day. Try to elaborate needs into statements of measurable results: "Human tellers take approximately two minutes to complete a routine transaction. The ATM should take no more than 45 seconds to complete routine transactions."

Also, expand on the requirements. Ask yourself, "Will they need software only on the ATM end, or will they need software to handle transactions and communications as well?" "What does the 24-hour-a-day requirement really mean?" "Will there be an interface to an ATM network such as Plus, Cirrus, or Star?" Then, review alternatives, asking such questions as "Should XYZ Bank use an IBM ATM or a Diebold ATM?"

Third, *restate this requirements specification in your words to the customer.* By doing this, you get the customer to focus on the details of what is required before a spec is written.

The fourth and final step in the process of writing a functional specification is to *begin your design spec process.* Once you've mapped out the System Requirements Specification—*what* is required to meet the needs—you may write a design spec describing the process of *how* it is to be done.

Guidelines for Writing Specifications

The following four guidelines may help you avoid the common pitfalls in writing specifications:

1. *Use clear technical language.* Rather than avoiding technical language, assume that the customer has technical staff to review technical portions of the spec. Watch out for formal language, legalese, or punctuation that invites ambiguity, sounds stilted, or changes the meaning of the description:

Unclear	*Better*
All ATM customer interface shall be displayed via the use of a 160-byte (4×40) gas plasma message-display device.	Messages to the ATM customer will be displayed on a four-line gas plasma display device.

2. *Be precise and concise.* State requirements as precisely as you understand and in a way that allows for testing once implemented. Use the fewest words to complete the descriptions. Two or three short sentences are preferable to one long sentence:

Unclear	*Better*
The updated data will be re-displayed within 5 seconds after the data is entered and validated.	The updated data will be re-displayed within 5 seconds after the customer has entered the data. The data will be validated by the system once it has been entered by the customer.

Although accuracy is vital in spec writing, readability is still important. By editing yourself and reducing your thoughts to the fewest possible words, you not only force yourself to be precise, but you can avoid multiple and often conflicting restatements of the same basic thought, and reduce the chance for ambiguity or misunderstanding.

3. *Be complete—but don't overdo it.* Technical staffers sometimes think that to be "comprehensive" they must research and write about all of the details or possibilities for a given requirement. But covering the important details doesn't imply covering everything (the "everything but the kitchen sink" approach that so many technical people take). Initially, concentrate on describing just the *normal* possibilities expected by the customer. This helps convey to the customer your understanding of the basic requirements and helps you create software using the descriptions appearing in the spec.

4. *Use the present tense.* We have seen a number of specifications that use *should* or *must* or the future tense *will* when describing what a proposed system will do in the future. Each word may be justified by particular circumstances, and, in those cases, we can only urge you to be consistent. But our preference is to keep specification descriptions in the present tense. For example:

"The CPU is composed of six assemblies."

"The CC monitors and controls the flow of signals."

"The System Controller performs the following functions."

As in most other writing, the present tense is dynamic and immediate. Once you start using it, you will be able to avoid the confusion of shifting from *should* to *will* to *is* to *must* that characterizes so many specifications.

7

Technical Articles, Papers, Abstracts, and Reports

Whether they work in academia or industry, in basic research or in applied technology, engineers and scientists must document their work and present their findings to others in articles, papers, and reports.

This chapter covers some of the organizational schemes and writing techniques used to prepare such materials.

Technical Articles and Papers

There are more than 6,000 business, technical, academic, scientific, and trade publications in the United States. They publish hundreds of thousands of technical articles and papers each year.

Technical publications are the medium through which engineers and scientists tell their peers in other organizations about their work. In academic journals, such as the *Bulletin of the Atomic Scientist* or the *Journal of the American Medical Association,* the authors are scientists reporting the results of their research to fellow scientists.

Trade journals, however, prefer a more practical approach. The articles in *Chemical Engineering, Machine Design, Elastomerics,* and other trade publications provide information that helps engineers and managers in industry do their jobs better.

Technical articles, then, are written *by* technical professionals *for* technical professionals. Why would a busy research scientist or plant engineer take the time and trouble to write for publication? We can think of at least six reasons:

1. Publishing offers personal satisfaction.
2. It increases the author's status as a technical expert.
3. It provides good publicity for the author's company.
4. It earns professional prestige, increasing the author's chances for tenure or promotion.
5. Writing teaches the author more about the subject.
6. The author is contributing to the pool of technical knowledge and helping others learn.

We live in an age in which the amount of information is growing at a frantic rate, and that's especially true of technical information: More and more papers and reports in all areas of science and technology are published each year.

Cambridge Scientific Abstracts, a publisher in Bethesda, Maryland, puts out volumes containing the abstracts of key papers and research reports in a variety of scientific and engineering fields. Table 7-1, which shows the number of papers and reports Cambridge abstracts in various fields each year, illustrates the profusion of scientific and technical literature.

Writing Technical Articles and Papers

Technical papers presenting the results of scientific research are written using a style and organizational structure similar to those of the research report, discussed later in this chapter.

Technical articles vary in style and technical depth, depending on the journal or magazine publishing the article. Many trade publications provide "author's guidelines" that dictate the tone,

Table 7-1. Abstracts of papers and reports published annually in selected scientific fields (source: Cambridge Scientific Abstracts, 1991–1992 catalog).

Topic:	Number of papers abstracted:
Algology, mycology, and protozoology	9,900
Amino acids, peptides, and proteins	9,900
Animal behavior	5,000
Aquatic pollution and environmental quality	6,000
Aquatic sciences and fisheries	34,100
Bacteriology	13,200
Biological membranes	8,800
Biotechnology research	4,800
Calcified tissue and metabolism	3,200
Chemoreception	1,500
Computers and information systems	17,600
Ecology	11,000
Electronics and communications	9,600
Entomology	8,800
Genetics	13,200
Human genome research	3,600
Immunology	13,200
Industrial and applied microbiology	7,700
Marine biotechnology	1,200
Mechanical engineering	9,600
Neuroscience and endocrinology	16,500
Nucleic acids	11,000
Oncogenes and growth factors	2,000
Pharmacology	4,000
Solid-state technology and superconductivity	8,000
Toxicology	7,700
Virology and AIDS	8,800

length, style, and format of articles; your best bet for preparing an article suitable for a particular publication is to consult the author's guidelines and review several back issues for style and content.

The technical expert is not expected to write with the flair and style of a professional writer. Journal editors will polish and rewrite the engineer's manuscript to make it more interesting and readable. But a well-organized, cleanly written first draft does increase your chances for publication; if your draft is *too* poorly written or difficult to understand, the editor may reject it.

Many corporations, including Raytheon, Westinghouse, and International Paper, publish their own magazines for distribution to customers and employees. The articles in these "house organs" are shorter, less technical, and newsier because many of the readers are nontechnical.

House organs usually generate stories from within the corporation. But trade publications are always on the lookout for new and exciting stories from industry and academia. Types of articles published include case histories, industry roundups, and stories on market trends, controversial issues, new products, improved technologies, ongoing research and development, new materials, new manufacturing techniques, energy-saving ideas and systems, environmental compliance, service, engineering, quality management, and performance.

If you would like to see your name in print, contact the editor of the journal for which you want to write and ask how to go about submitting an article. Although you won't gain the fame and fortune of a Stephen King or a Sidney Sheldon, technical publications have their own rewards.

Abstracts

At a recent meeting of the American Institute of Chemical Engineers (AIChE), more than 300 technical papers were presented on topics ranging from "Exposure of Steel Containers to an

External Fire" to "Thermodynamic Availability Analysis in the Synthesis of Energy-Optimum and Minimum-Cost Heat Exchanger Networks."

Prior to the meeting, each AIChE member was mailed a 94-page program containing abstracts of all the papers. By scanning abstracts, meeting participants could quickly decide which presentations to attend.

An abstract is a short (generally 150 words or less) statement of the contents of a report, paper, or other document. The abstract introduces the subject matter, tells what was done, and presents selected results.

Your readers cannot possibly read all the reports and papers that come across their desks, but most will at least scan an abstract to see whether your report is of interest to them. A well-written abstract is the best means of convincing the right people to read your report.

Abstracts of thousands of technical papers presented each year are permanently collected for future reference in volumes such as *Science Abstracts, Abstracts on Hygiene, Applied Mechanical Reviews*, and *Human Genome Abstracts*. Some abstract collections are available on CD-ROM and magnetic tape databases.

Many technical writers are employed full-time to edit and index abstracts for these publications. (Before turning to science fiction, Arthur C. Clarke, author of *2001: A Space Odyssey*, was an assistant editor of *Science Abstracts*.)

Abstracts describe complex scientific research in fewer words than are on this page. Therefore, they must be extremely concise.

Here is a well-written abstract from a paper presented at a conference of pulp and paper engineers. It tells the whole story in four fact-filled sentences:

Energy Requirements for Coated and Uncoated Papers

A study was made of the energy required to produce coated and uncoated papers. Energy use requirements were determined for each process from pulping through coating. Raw material energy needs were included in the study. Based on the total energy concept, it was determined that the production of coated papers

requires 5 to 12 percent less energy than the production of uncoated papers of similar weight and brightness.*

The abstract clearly states the important result in quantitative
terms: Producing coated paper takes 5 to 12 percent less energy
than producing uncoated paper. However, we should have been
told *how* energy-use requirements were determined and *why* the
production of coated paper consumes less energy; a good abstract says how the work was done and why (if known) a particular result was achieved.

In the abstract below, the work done, techniques used, and
limits of precision are described in one tightly written
paragraph:

A Precise Optical Instrumentation Radar

The instrumentation tracker described provides real-time posi-
tioned data on high-speed cooperative targets with precisions of
± 1 m at ranges between 300 m and 10 km. Unambiguous range
is determined by a precise digital FM-CW ranging technique at
a rate of 15 per second. A target-mounted beacon and a narrow
laser-ranging beam permit measurement of target position to
values much less than the target dimension. Azimuth and ele-
vation angles are read out by precision shaft angle encoders and
recorded in binary form, along with range and time, on a magnetic tape or directly into a real-time computer.†

Although what goes into a particular abstract depends largely
on what is in the report—for example, you can't summarize
conclusions in your abstract if the researcher makes no conclusions in his or her report—most abstracts include the following
items:

---

*G. M. Heim and R. L. Lower, "Energy Requirements for Coated and Uncoated
Papers," excerpted in *Instructions for Paper Preparation and Presentation at
TAPPI Meetings* (Atlanta: Techical Association of the Pulp and Paper Industry,
1981), p. 19. Copyright © 1981, TAPPI. Reprinted with permission.
†T. C. Hutchison, A. A. Hagen, H. Laudon, and C. R. Miller, "A Precise Optical
Instrumentation Radar," *IEEE Transactions on Aerospace & Electronic Systems*
AES-2, no. 2 (March 1966).

- A statement of the problem. This should be one sentence, if possible.
- A brief discussion about the approach taken to solve the problem.
- The *main* result. Keep other accomplishments out of the abstract.

Reports

Technical reports are the documents in which engineers, scientists, and managers transmit the results of their research, field work, and other activities to people in their organization. Here's what the University of Rochester's department of chemical engineering has to say about engineers and report writing:

> The importance of being able to write a good report cannot be emphasized too strongly. The chemical engineer who carries out an investigation or study has not completed his job until he has submitted a report on the project. The true value of the project and the abilities of the investigator may be distorted or unrecognized unless the engineer is able to write a commendable report.

Often, a written report is the only tangible product of hundreds of hours of work. Rightly or wrongly, the quality and worth of that work are judged by the quality of the written report—its clarity, organization, and content. Therefore, it pays to take the time to write a good report.

Technical professionals produce a number of different types of reports, summarized in Table 7-2.

Elements of the Technical Report

The report most frequently written by engineers, research scientists, and other technical professionals—as well as by science and engineering students—is the research or lab report.

Although research reports can take many forms, most contain the following major sections: cover and title page, abstract, table

Table 7-2. Types of reports.

Type	Description and purpose
Periodic report	Report submitted at regular intervals to provide information on the activities or status of the organization. Bank statements, annual reports, and call reports are examples of periodic reports.
Progress report	Update on an ongoing activity as it is being carried out. The activity may be construction, expansion, research and development, production, or other projects.
Research report	Results of research, studies, and experiments conducted in the lab or in the field.
Field report	Results of an on-site inspection or evaluation of some field activity, which might be construction, pilot-plant tests, or equipment installation and setup.
Recommendation report	Report submitted to management as the basis for decisions or actions. It makes recommendations on such subjects as whether to fund a research program, launch a project, develop a new product, buy a piece of capital equipment, or acquire a company or technology.
Feasibility report	Report that explores the feasibility of undertaking a particular project, venture, or commitment. It examines and compares alternatives, analyzes the pros and cons, and suggests which, if any, of the alternatives are feasible.

of contents, summary, introduction, body, results, conclusions and recommendations, nomenclature, references, and appendix. Let's take a brief look at each of these sections.

1. *Cover and title page.* The cover and title page create the reader's first impression of your report. The cover should be cleanly typed but not gaudy; do not try to fancy it up with borders, stars, or similar treatments. The title page should be neat; the title should tell the reader exactly what the report is about.

2. *Abstract.* The abstract is an informative, concise, one-paragraph statement of the work performed, its objectives and scope, and the major conclusions reached. Several sample abstracts were provided earlier in this chapter.

3. *Table of contents.* The table of contents lists every section heading and subheading and the page number on which they appear. Tables, figures, charts, and graphs are listed separately at the end.

4. *Summary.* While the abstract gives the reader enough information to decide whether to read the report, the summary presents its entire contents in a few hundred words—usually one page or less. It covers the purpose of the work, the goal, the scientific or commercial objective, what was done, how it was done, and the key results. Summaries are usually not included in shorter reports (those of fewer than 25 pages).

5. *Introduction.* The introduction tells readers—including those not familiar with the subject matter or the reason for writing the report—the purpose of the report. It provides background material, theory, and explanation of why the work was done and what it accomplished.

The introduction should:

- present the nature and scope of the investigated problem
- put into perspective the importance of the research as it relates to scientific knowledge or commercial operations
- discuss findings from previous research and other pertinent literature, if such material exists
- state the method of investigation
- present the key results of the research.

Here's an example of a good introduction:

Marketing had requested R&D to research conditions under which E-Z Bond dentifrice formulations could prove harmful. In response, R&D conducted acute oral toxicity tests on the current dentifrice formulation, which contains 0.8% NaMFP, and two experimental dentifrice formulations containing 1.6% and 2.4% NaMFP, respectively.

Each of the dentifrice formulations was administered by oral intubation into male rats at a dosage of 30 grams per kg of body weight, then to three groups of female rats at dosages of 30, 15, and 10 grams per kg of body weight.

Subjects were evaluated for appearance and behavior, weight loss, mortality, and gross pathology at necropsy. No mortality was observed in any of the tests.

6. *Body.* The body of the report contains the detailed theory behind the work. It also outlines the apparatus and procedures used, so other researchers can follow these steps and repeat the experiment.

7. *Results.* This section presents experimental data, observations, and results, along with a discussion of the meaning, significance, importance, and application of these results. The discussion should present the principles, relationships, and generalizations supported by the results; point out any exceptions or lack of correlation; explain (if possible) why such exceptions or deviations occurred; and show how the results compare with results achieved by other researchers.

Here's an excerpt from the results section of a report:

Results from this investigation show that the toxicity of the dentifrice formulations in males was directly related to the fluoride concentration. Mortality data based on the level of fluoride administered is as follows . . .

When the doses administered to females were adjusted to provide 31.7 F/kg of body weight from each of the three formulations, no mortality was observed.

8. *Conclusions and recommendations.* The conclusions are a series of numbered statements showing how the results answered questions raised in the stated purpose of the research. On the basis of the results and conclusions, the researcher can make recommendations about whether further research is needed or how the results can be applied commercially.

9. *Nomenclature.* This section lists, in alphabetical order, the symbols used in the report and the proper unit of measure for each.

10. *References.* A research report includes an alphabetical bibliography listing the technical literature (books, papers, brochures, reports, conference proceedings) used by the author as secondary research and reference material.

11. *Appendixes.* Appendixes contain any sample calculations, tables, mathematical derivations, sets of measurements, calibration data, or computer printouts that are too long, cumbersome, or unimportant to be included in the main body of the report.

These sections are typical, but the order may vary, and one or more may be omitted, depending on the length and complexity of the report. Also, many companies require reports to conform to a particular format or style. If you have to write a report, check with your supervisor or technical publications department to see whether your organization has guidelines for technical reports.

8

Letters and Memos

Letters

In addition to routine correspondence, there are several different kinds of letters technical professionals have to write, such as:

letters of technical information
letters of transmittal for reports and proposals
letters of instruction
announcements of new products, facilities, policies, and services
answers to product inquiries

Years ago, business letters were often written in formal language that sounds stiff today. Now, technical communications are less stuffy and more personal. A letter is a personal communication from one human being to another—not from one faceless department or organization to the next. Effective letters get their messages across by being friendly and helpful.

Writing in a conversational tone doesn't mean that the technical writer needs to boil down everything he or she writes to the lowest common denominator of understandability. If you are writing to a technician, you may use as many technical terms as your reader understands. All we are saying is that there is a tendency to string together big words, long sentences, complex thoughts, and long paragraphs. Often, the technical writer, carried along on this path, forgets that the reader is a human being who appreciates the same conversational flow, informality, and smoothness that the recipient of a business letter would appreciate.

Writing in an easygoing, conversational tone takes practice. As a start, eliminate stilted expressions from your writing:

Instead of:	*Use:*
Enclosed please find	Enclosed is
Attached please find	Attached is
I am forwarding herewith	I'm sending
Per your inquiry of	In response to your question
If you will kindly inform us	Please inform us
In accordance with your wishes	As you requested
Awaiting your earliest reply, we remain	Sincerely
In view of the above reasons	For these reasons
You will please note that	(delete)
Please don't hesitate to call	Please call me
Pursuant to your request	At your request
As per our conversation	As we discussed
The undersigned	I
Herein	In this letter
Above stated	(delete)
Above referenced	(delete)

The two letters below show the difference between old-fashioned corporatese and today's more personal approach. Both are answers to an inquiry about the company's product:

SPARTAN CO., INC.
770 LEXINGTON AVENUE
NEW YORK, N.Y. 10021

October 23, 1994

Mr. Ron Brick
Chief Engineer
Chemtech Corp.
130 Sumner Ave.
Akron, Ohio 44309

Dear Mr. Brick:

CHEMICAL PROCESSING magazine has informed us of your interest in the level detectors manufactured by our firm for use in the chemical processing industry. As you may perhaps know, we are one of the oldest and most well-respected manufacturers of such equipment, and our product line includes the following types of level detectors: Beam Breaker, Bubble, Diaphragm, Capacitance, Conductive, Differential Pressure, Displacer, Float and Float and Tape, Glass and Magnetic Gauge, Hydrostatic Pressure, Inductive, Infrared, Microwave, Optic Sensor, Paddle, Pressure-Sensitive, R-F Admittance, Radiation, Sonic Echo, Strain Gauge, Thermal, Tilt, Vibration, and Weight and Cable level detectors that are described in the enclosed technical sales literature.

Since you may also have requirements for our other types of process equipment, we are enclosing our All-Line Catalog and Data Sheets with the request that you fill in the Data Sheets with as much information as you have available, returning them to us for the consideration and recommendations of our Engineering Department, enabling us to quote you, if possible, on specific applications. Finally, as our Company is now in its fourth decade of continuous service to its many Customers in this Country and Abroad, we are sending along a reprint of our latest annual report which will give you more information on our activities. We will await with interest your specific inquiries. Thank you once again for contacting us.

Very truly yours,

John N. Guterl, President

MARS MINERAL CORP.
P.O. BOX 128 • VALENCIA, PENNSYLVANIA 16059
412-898-1551 TELEX 866452

September 18, 1994

Mr. L. Moore, Proj. Engr.
Spartan Co.
770 Lexington Ave.
New York, NY 10021

Subject: Pelletizing Information

Dear Mr. Moore,

Thanks for your interest in our Pelletizers. Literature is enclosed which will give you a pretty good idea of the simplicity of our equipment and the rugged, trouble-free construction.

The key question, of course, is the cost for equipment to handle the volume required at your plant. Since the capacity of our Pelletizers will vary slightly with the particulates involved, we'll be glad to take a look at a random 5-gallon sample of your material. We'll evaluate it and get back to you with our equipment recommendation. If you will note with your sample the size pellets you prefer and the volume you wish to handle, we can give you an estimate of the cost involved.

From this point on we can do an exploratory pelletizing test, a full day's test run or we will rent you a production machine with an option to purchase. You can see for yourself how efficiently it works and how easy it is to use. Of course the equipment can be purchased outright too.

Thanks again for your interest. We'll be happy to answer any questions for you. Simply phone or write.

Very truly yours,

MARS MINERAL CORPORATION

Robert G. Hinkle
Vice President, Sales

The letter from the Spartan Company is fairly straightforward and comprehensible. But would you ever talk to a customer in such long-winded sentences? (The second sentence in the first paragraph is 83 words long.) The letter is full of letter writers' clichés ("... has informed us of your interest...," "We will await with interest your specific inquiries"). Words such as *company, country,* and *abroad* are capitalized for no reason. Plus, this letter has "form letter" written all over it, since there is nothing in it that seems custom-written to the prospective customer.

Mars Mineral Corp. does much better with its livelier, more lucid reply letter. The letter is friendlier, the paragraphs and sentences are shorter, and the tone is more conversational—like one friend talking to another. While the letter from the Spartan Company merely repeats information found in the bulletins mailed with the letter, Mars Mineral suggests a course of action—sending in a material sample for evaluation—that can solve the customer's problem and lead to the sale of Mars Mineral's pelletizer.

Below are examples of other typical technical letters, including a letter of technical information, a letter of transmittal, a letter of instruction, and an announcement of a new product:

Letter of Technical Information

September 29, 1992

Ms. Ruth Kanulen
Barry Chemical Co.
234 Victoria Street
New Octavia, KS 54678

Dear Ms. Kanulen:

At your request, I am sending 1.3 Kg 0.2% AGE-modified amioca starch and a Material Safety Data Sheet for your research program.

As I suggested, you should study the effect of AGE on the starch backbone and molecular weight of starch.

Natural Polymer Research will prepare these starches for you.

Please keep me updated on the progress of the starch/graft research.

Regards,

Bonnie Bonnard

Letter of Transmittal

January 1, 1993

Mr. Bernie Segal
Laboratory Technician
Commerce Chemical Co.
1233 New Trenton Street
Floddert, VA 88898

Dear Mr. Segal:

Enclosed is a technical service report and a laboratory procedure for emulsification of FO-BRAN 55 on a small scale.

The report describes some of FO-BRAN's physical properties, gives examples of sizing results in the field, and explains field emulsification procedure. The laboratory procedure tells you how FO-BRAN emulsion is prepared in smaller amounts.

I hope the information enclosed is sufficient to introduce you to FO-BRAN and allow you to run your evaluations successfully.

If you have any questions, please call me at (718) 999-0918.

Sincerely,

Robin Deere
Paper Development Specialist

Enclosures

Letter of Instruction

June 4, 1991

Ms. Kathryn Lukens
Dirctor—Consumer Information Center
American Rail Corp.
1330 North R Street
Tacoma, WA. 89898

Dear Ms. Lukens:

This will confirm my telephone instructions to Mr. George
Hotchkiss concerning the diversion of ASDS 87778.

This car, shipped from North Washington, Delaware, by Star
Chemical Company via AmRail direct, was originally consigned to
ABC Chemicals Corporation at Nutley, New Jersey. Please arrange to
divert ASDS 87778 to Monmouth Chemical Company at Kearney,
New Jersey, via AmRail direct. Any charges connected with this
diversion should be sent to me.

Sincerely,

Dr. Frederick Loosey
Senior Transportation Analyst

Announcement of New Product

Ms. D. Lee
Director of Operations
Kooper Air Terminal
PO Box 343
Fargo, ND 45464

Dear Ms. Lee:

I am pleased to enclosed a "first customer" GateProgram software license and payment plan for Kooper.

Included in this license are our software warranty, maintenance, and extended warranty agreements.

There's no other product of its kind on the market today that has proven to be as effective in the managing of gate operations of a major airline. The installation, training process, and guidelines of support will allow Kooper to perform without the purchase of an additional site license.

If the package is agreeable, please call me to discuss setting up your contract.

Sincerely,

Sam Rabinowitz
Manager, Software Marketing

Memos

Letters are usually written to those outside your organization; memos (short for *memoranda*) are usually sent internally. And, while technical people have a habit of making their external communications too stuffy, they sometimes make their memos to colleagues unnecessarily brusque.

For example, look at the following memo:

To: Programmers

From: T. Gray

Due to the poor response in filing off documents, the system is still working at a high capacity. Although it had been urged that filing off be done regularly, it appears that no one has followed the correct procedure. Therefore, the system is working harder, and there have been several reports of malfunctions.

In view of this, Maria Jones, Systems Supervisor, has asked me to advise you that the system will be reinitialized on Wednesday, June 22. In other words, the system will be cleaned out. This will result in a shutdown of about 2 hours.

You must file off all your documents no later than Tuesday, June 21. As the reinitialization process deletes anything left in the system's disk, we will assume that you do not need any documents remaining in the system on June 22.

Thank you for your cooperation.

This memo is sometimes brusque—even offhandedly caustic. The emphasis is on the poor response in filing off documents; subtle sarcasm ("It appears that no one followed the correct procedure") doesn't motivate people to change. The prescriptive tone is topped off by a veiled threat: Either remove your documents before June 22, or "we will assume that you do not need" them!

This person put anger first, and therefore we don't learn what the memo is about (the reinitialization) until the second paragraph. The final paragraph, a type of thanking in advance, is ludicrous since the people have not cooperated. Thanking them for something they haven't done won't make them do it.

Here's a better way of writing the memo:

> On Wednesday, June 22, the system will be shut down for two hours to be reinitialized. Therefore, you must file off your documents no later than June 21. Files remaining on the systems after that date will be purged.
>
> This reinitialization is needed because the system has been working at a high capacity as a result of people failing to file off regularly. Once again, here's the procedure for filing off documents:

Our memo steers away from suppressed hostility; it presents facts.

Tips on Writing Letters and Memos

There are too many kinds of letters and memos written in the life of a business or technical writer for us to cover all situations. The following nine tips on writing letters and memos can be applied widely:

1. *Use a person's name and title whenever possible.* People like their names used and spelled correctly. Usually, a phone call to the person's office or organization will give you that information.

If you can't get the person's name, then use a generic title to address your reader. For example, if you're writing to a department store for credit, write to "Dear Credit Manager." Avoid vague salutations in letters such as "To Whom It May Concern" or "Dear Sir/Madam." If you send your letter to someone specific, the letter is more likely to be taken seriously and not passed around from person to person than if you address the letter vaguely.

Finally, if you're addressing a wide variety of people, use a generic title such as "Dear Reader," "Dear Technical-Services Managers," and the like.

If you know the name but not the sex of the reader, we suggest you use the full name in the salutation without identifying the sex (e.g., "Dear Terry Smith" or "Dear Leslie Fahrenkrug").

2. *Use a* Re *line to signal the subject matter of the memo.* The *Re* line is an optional separate line above the salutation and

below the reader's address. *Re* lines can be useful in both letters and memos to highlight key details for quick reference, such as relevant invoice numbers, dates of past conversations or correspondence, file numbers, or client names.

Re lines also orient the reader as to the content and nature of the communication ("Re: Your upcoming performance review of 1/4/94"). It's especially helpful to remind readers that you are writing about a topic on which you've had previous communication ("Re: September reorganization of ORACLE database") or that you're responding to their letter ("Re: Your memo of 9/2/93").

Re lines hold information you might otherwise be forced to place in the opening sentences of your letter or memo. If you use a *Re* line, you can avoid such awkward opening sentences as "In response to your August 2d letter regarding Ed and Martha Colby's account . . ." Just put the date of the letter, the names, and the account number in the *Re* line:

Re: Response to August 2d letter about Ed and Martha Colby's account

Then start, "After researching the missing interest in the Colbys' account, I found . . ."

In memos, the *Re* line is usually placed two lines directly beneath the date. Here are three examples of succinct *Re* lnes:

Re: Need to Improve Product Movement between Plants

Re: Weekly Progress—August 16, 1991

Re: Expectorant pH Values of Tartar-Control Pastes

3. *Give your reason for writing in the first paragraph.* As in all other forms of writing, letter and memo writing requires you to be organized. If you put too many details in the first paragraph or fail to get to the point of your message, your reader may not stay with the letter long enough to find the important information.

Certainly, there are exceptions to this rule. You might not do it in a letter or memo to a subordinate. Nor would you do it when you wanted to use your first paragraph to thank someone for doing something or to bridge a gap from an earlier communication. The important issue is to avoid meandering into your idea so that you fail to get to the point, or get to it three or four paragraphs into the message.

4. *Establish an appropriate order for your responses.* If your letter or memo is in response to an earlier communication, you should answer any questions that were asked—but not necessarily in the same order. Although generally it's a good idea to answer questions in the order they're brought up, recognize that some letter writers ask questions in no particular sequence other than the order in which they occur to the writer. In these situations, you must take control by organizing your responses according to your own logic. And if you can anticipate the kind of information your reader will need, you'll spare both your reader and yourself the need for additional correspondence.

5. *Use the proper format.* Modern letter writing follows one of three basic formats: block, modified block, and semiblock.

In block style, all lines start flush with the left margin.

January 1, 1992

Mr. A. Smith
Research and Development Director
AmTar Corporation, Inc.
123 Home Street
New York, NY 10001

Dear Mr. Smith:

I enjoyed seeing you last week at the convention. Here are the brochures you requested.

Sincerely,
Terry Kirkland
Marketing Director

In the modified block style, the date, *Re* line, closing, and signature are to the right of center; all other lines start flush left. Paragraphs are not indented but are separated by a line space.

September 12, 1992

Ms. Michelle Bailey
Sales Manager
ART Services, Inc.
121 Shiretown Street
Lawnview, CA 91112

Re: Zonarek 7116 resin

Dear Ms. Bailey:

Confirming our August 27th conversation, we must bring your price of Zonarek 7116 polyterpene resin to $0.76/lb. The price will be effective October 1, 1992.

Please send me acceptance of this in writing.

Sincerely,

John Reese
National Sales
Manager

In the semiblock style, the date, *Re* line, closing, and signature are to the right of center; all paragraphs in the body of the letter are indented, usually five spaces.

October 5, 1991

Mr. Patrick Lawson
Lawson, Claven, and Peterson
One Blue Mountain Plaza
Pearl River, NY 12009

Re: Sewer Flow Meter
Project No. 88-90

Dear Mr. Lawson:

As we discussed, based on city water consumption data, the flow meter does not appear to work properly. At your request, we transmit the following data:

1. Average city water consumption
2. Total suspended solids

Please submit your proposal to analyze the systems and calibrate the flow meter.

Sincerely,

Nicholas Tortelli
Project Engineer
Facilities
Engineering

Often, a company or department prefers one of these styles to the others; if not, choose your style—people who use word processors are fond of block style, since word processors have programmed margins—and stick with it in all your correspondence.

As to the rules for spacing, leave three spaces between the date and the inside address; leave two spaces between the end of the address and the salutation; leave two spaces between the salutation and the start of your letter; leave two spaces between the end of your letter and the closing; leave four spaces between the closing and your typed name (enough room for your signature).

In memos, double-space the *To, From, Subject,* and *Date* lines. Single-space the memo. If you use block form, you will not indent the paragraphs. So, to make sure the paragraphs don't look too chunky, keep them short (usually fewer than eight lines each).

6. *Keep letters and memos brief.* People neither expect nor want long letters and memos. If possible, limit letters to one or two pages. The same goes for sentences and paragraphs: Keep them short and to the point. (We recommend limiting the first paragraph to one or two sentences; subsequent paragraphs can be longer.) If you have a lot to say, you can extend your letter to two pages. You have other options, too: Enclose a separate report containing the statistical or factual details, saving your letter for the highlights of the message. Too often, people refuse to read letters that are more than a page long—they lose interest.

7. *End your letter or memo by telling the reader what happens next.* Whether you need a response by a certain date, want someone's approval on something, or just want the recipient to know what you'll do next, let your reader know in the closing sentence. For example, "Please sign the form and send it to me by January 6" and "I'll give you a call on Tuesday to follow up" are specific closing sentences.

These sentences avoid the clichéd and vague "Thanking you in advance" (Don't thank people for doing something they haven't done—or agreed to do) or "If you have any questions, please feel free to contact me" statements.

It's all right to end on a vague note occasionally, since we don't always have specific actions and firm commitments upon which to comment. When just keeping in touch or being cordial, there's nothing wrong with ending with "I look forward to seeing you soon." But don't use a close like "If you have any questions . . ." when the letter or memo you've written is so straightforward and simple that one would probably *not* have questions.

8. *Making the closing simple.* Memos don't require a formal closing, but letters do. Endings such as "Very truly yours," "Cordially," "I remain," and "Kind regards" should be avoided because they are old-fashioned, stiff, and servile. "Sincerely" works

well. If you have a personal relationship with the reader, then "Regards" is acceptable.

9. *Adopt an easy-to-read format.* Use wide (minimum 1¼ inch) margins. In lengthy letters and memos, use subheads to break the text. Use underlining to emphasize subheads or other important information. Use bullet points or numbers to list a series of items. A handwritten postscript can add immediacy to your message while capturing attention. For a crisp format, make sure your typed or word-processed text is not close to or in conflict with any graphic elements of the letterhead design.

A Word about E-Mail

In the past five years, many people in the technical community have become used to transmitting information within their organizations by E-mail (electronic mail), an informal way of sending messages between computer terminals.

E-mail is now widely available, capable of reaching destinations worldwide often at less cost than a telephone call or even postage. Its origins in the research and academic communities, combined with its rapid rise in popularity, have led to technical standards for how computers are to address and deliver E-mail, but standards for content and style have been left up to each sender. Following are a few guidelines for good E-mail use:

1. *Keep it short.* Short messages are appropriate for E-mail. Lengthy documents sent unsolicited to numerous recipients are inappropriate. It's better to send a message announcing the availability of the document than the document itself.

2. *Be consistent.* Since E-mail is so informal, there is a tendency to write in short, staccato sentences and phrases; to keep the message in all capital letters; and generally to ignore the rules of punctuation and spacing. Appearance still counts. Treat your E-mail the same as any other professional communication.

3. *Restrain the urge to "return fire."* The immediacy of E-mail may make the recipient feel compelled to compose an immediate response. Don't be hasty, especially if you are responding to an E-mail message that has made you angry or defensive.

4. *Be sure to refer to the original message.* For example, if you receive a meeting notice, mention it in your reply: Say "I will attend the meeting scheduled for . . ." or "Re: Your E-mail message of 9/10/92, 9:08 am . . ." instead of "I will attend" or "Yes, I'll be there."

5. *Paraphrase when necessary.* Although E-mail started out as a handy way to send short messages, some people "piggyback" new messages on top of the original messages, often causing confusion.

Since E-mail has a feature that allows the recipient of a message to add some comments and forward it, a short message can become a lengthy chain of comments. Rather than perpetuate the chain of messages, paraphrase the comments and original message. A sample paraphrase: "I received a meeting notice for August 12, 1992, at 9 am, but John, Joe, and Mary all said they can't attend, so please reschedule it."

9

Manuals and Documentation

> *If any single force is destined to impede man's mastery of the computer, it will be the manual that tries to teach him how to master it.*
>
> William Zinsser, *On Writing with a Word Processor*

Types of Manuals

Many large engineering firms have departments that produce manuals about the company's products. Most software companies use staff or outside technical writers to prepare manuals for their products. Systems development professionals are often responsible for creating the documentation that goes along with the systems they produce.

Table 9.1 lists just a few of the many types of manuals technical writers are called upon to write. These may require separate volumes or may appear as sections of a single volume.

What makes for an effective manual? In an article in *Training and Development* (March 1992), Mona J. Casady, a professor in the department of administrative office systems at Southwest Missouri State University, says effective manuals have the following characteristics:

- they are well written
- they are attractively designed
- their format makes it easy for users to follow instructions accurately

Table 9-1. Common types of manuals.

Type	What It Tells
Installation manual	How to install a device or piece of equipment correctly and safely. Typically includes wiring diagrams and exploded views to aid in assembly.
Instruction manual	How to operate equipment. Usually shows control panels with call-outs indicating the function of each control or readout.
Operations manual	How and why a piece of equipment works in theory and practice. Typically includes schematic diagrams, blueprints, tables of operating data, performance curves, and specifications.
Sales manual	How to specify and purchase equipment. These manuals contain product specifications, pricing, and other information salespeople use to sell products.
System documentation	How a system was designed, what it does, what the components are.
User's manual	How to use software, a computer, or any other system or device. Often illustrated with pictures of screens.

- the format enables designers to revise easily
- they are illustrated appropriately to enhance understanding
- they help people do their work correctly, efficiently, uniformly, and comfortably.

Guidelines to Help You Write Better Manuals

In addition to the rules of clear technical communication presented in chapters 1 through 5, the following guidelines can help you write manuals that are easy to read and easy to follow.

1. *Remember that manual writing is instruction writing.* Like a well-written cookbook, a manual does nothing more than present instructions for completing a certain task. The only difference is that the operations described in technical manuals are often highly complex and difficult to understand.

You will be a more effective writer of manuals if you keep in mind that your mission is to *give instruction*, not sell, impress, or dazzle your audience with technical knowledge or prose style. Almost everyone dislikes reading manuals, so the easier you can make things for your reader, the better.

The best way to improve your ability to write clear manuals is to practice giving instructions. If you have trouble giving clear directions to car drivers, for example, you'll probably have trouble explaining how to use CAD/CAM software or install a programmable controller.

Develop your ability to give clear instructions by writing instructions for nontechnical activities. For instance, you might try writing instructions on how to put on a necktie, then see if your 9-year-old son can follow them without help from you.

Writing usable instructions for seemingly simple acts can be quite difficult. But once you master it, you'll find writing instructions for technical activities is not much different—and not much more difficult.

2. *Be complete.* A cookbook gives this recipe for chicken soup:

In a large kettle, bring 6 quarts of water to boil. Add a clean chicken and simmer for 30 to 60 minutes. Then add vegetables, cover the kettle, and simmer for another 60 minutes. When the soup is done, remove the chicken and all the vegetables except the carrots. Serve with rice or noodles.*

Our Favorite Recipes, Women's American ORT, New Jersey Dist. III (1974), p. 28.

The problem with recipes such as this is they assume a familiarity with cooking and kitchen procedures many readers do not possess.

For instance, we are told to place a "cleaned chicken" in the kettle, but the recipe doesn't say what size or kind of a chicken. A 6-pound chicken? a 3-pound chicken? a whole chicken? chicken parts? a fryer? a broiler? an oven stuffer roaster?

Also, what does it mean by "cleaned"? Aren't chickens sold in supermarkets already clean? You don't wash package steaks or chopped meat you buy from the supermarket. Does the author want us to wash the chicken?

The instructions say to "simmer for 30 to 60 minutes." That's a big range. How do you know when the soup is ready for the next step, the addition of the vegetables? By sight? taste? texture?

You may say I am nitpicking, but a person who has never prepared homemade soup before will have precisely these questions and may be intimidated by lack of knowledge and clear directions—so much so that he or she decides not to attempt cooking the soup. In the same way, a reader who finds your directions for using a computer or installing a chlorinator incomplete may return your product or at the very least be angry about the omissions.

Make sure you give complete directions. It is better to assume too little knowledge, experience, and familiarity with your technology on the part of the reader than to assume too much; we have never heard of any person complaining that a manual was too easy to follow.

Of course, you must assume *some* knowledge on the part of the reader, but where do you draw the line? Visualize your target audience, and write for that audience so the *least experienced or skilled member* of the group can understand what you are saying and follow your directions.

At the same time, there are some things—facts, terms, technical concepts, and so on—it is reasonable to assume your target audience knows. For example, it's reasonable to assume electricians understand the concepts of resistance and voltage, or that they know how to splice wire, and so you need not explain these things in your manual.

When in doubt, it's a good idea to state explicitly in the preface or introduction of your manual who the intended audience is and what knowledge you assume on their part:

This manual is intended to be used by technicians in installing the new Direct Residual Controller on existing chlorinator units. We assume the reader is familiar with chlorinator operations as well as the general principles of direct residual control.

This manual assumes a basic grasp of computer programming and system design. No prior knowledge of object-oriented design is assumed or required.

3. *Be clear and correct.* The technical writer must strive to make manuals clear, direct, and easy to follow. Few things are as frustrating to the owner of a product than a manual he or she can't follow or one that contains errors—for example, an instruction sheet for a piece of furniture that says there are screw holes where there aren't any.

Even difficult procedures can be made clear through careful writing. In his book *How to Dissect: Exploring with Probe and Scalpel* (New York: Sentinel, 1961, p. 105), William Berman approaches the complex operation of dissecting a frog in the same simple style as the chicken soup recipe. He doesn't make the instructions needlessly complicated simply because the subject is technical:

The heart is encased in a thin sac called the *pericardial sac*. Cut through the thin membrane of this sac with the tip of a very sharp razor. Do not cut the heart itself. Then spread the membrane with forceps to expose the heart.

Although the procedure is complex, the language is plain, conversational, and human. Reading this manual is like having a patient tutor at your side, helping you step by step to perform the operation described. And that's how all good instruction manuals should be written.

4. *Be unambiguous.* It is better to be repetitious and perfectly clear than brief and possibly unclear.

While it's important to make your manual as concise as possible, it's even more important to give instructions that the reader can follow with confidence.

Ambiguity in instructional materials makes readers uncomfortable and nervous. They want to be *sure* they're doing things right. If you feel your reader might be confused or uncertain about a particular task or procedure, repeating your instruction two or three times in different ways can help eliminate uncertainty and give the reader confidence.

Not only is a good manual written so that it can be easily understood; ideally, it is written so the instructions cannot be *mis*understood. And that's sometimes a tall order to fill.

William Berman, in his dissection instructions, says, "Cut through the thin membrane of [the pericardial] sac with the tip of a very sharp razor." Although his instructions are *cut the membrane*, he wants to write so that he cannot be misunderstood, so he adds, "Do not cut the heart itself." Although this may seem redundant—after all, if you're following step-by-step directions and the directions don't instruct you to cut the heart, you wouldn't do it—the author knows that cutting the heart is a common mistake students make, and so he emphasizes this point to be completely unambiguous.

Be unambiguous. Write not only to be understood, but also so you cannot be misunderstood.

5. *Use warnings.* Instruction writers are responsible for telling their readers what to do, but they must also take responsibility for telling their readers what *not* to do.

If performing a certain task might accidentally erase a harddisk file, give the operator an electrical shock, expose workers to toxic fumes, shut down a production line, or damage equipment, you must explicitly warn the reader against taking that action.

Warnings should be highlighted by setting them in boldface type, in all capital letters, in a large typeface, or with some other graphic technique. Putting a box around the warning is a good way to draw attention to it, because people always read text in boxes. Therefore, you can simply insert the warning into your document at the appropriate place, then make it stand out from the rest of the copy by drawing a box around it:

> WARNING: The XT-5000 chlorinator is designed for industrial and municipal use only and should not be used in swimming pools, parks, or for other recreational purposes.

> WARNING: The X-9 Control Unit should be unplugged from its power source before the front panel is removed and internal components are serviced. Otherwise, a severe electrical shock may result.

> WARNING: The reagent must be handled with extreme care because it contains sulfuric acid, a compound that can cause severe burns when it comes in contact with human skin. The technician should wear goggles and heavy rubber gloves when pouring the reagent, and the compounds should be mixed under a hood to prevent accidental inhalation of fumes. The lab area where reagents are handled should be equipped with an emergency shower and eyewash station within easy reach of the fume hood.

Boxed warnings should be using sparingly and reserved for only the most important warnings. Overusing the box or any other graphic technique makes it less effective. (For example, underlining a word on a page draws attention to it. But if you underline every other word, then no word stands out.)

Secondary warnings—those instructions you want to emphasize because of their importance but that do not warrant the box treatment—can be put in italics, underlined, or emphasized using some other graphic convention. You can also reinforce the importance of a warning or instruction by preceding it with the words *make sure*, as in these examples:

Before applying power, make sure the power supply in the equipment conforms to the form of primary power available.

Before closing the panel, make sure the screws are tight.

Make sure the set point does not exceed safe levels.

6. *Use the imperative voice.* Instruction manuals are written in the imperative mode.

Be direct. It's better to write "Connect the communications line" than the weaker, more passive "The communications line should be connected." As shown in the example below, effective instructions tell the reader what to do in the simplest, most direct language possible:

> Connect one end of the line cord with J3 on the back panel. This cord includes a ground wire for plugging into a grounded outlet. Before applying power, make sure the power supply in the equipment conforms to the form of primary power available. Required fuses are listed below.

When writing in the imperative mode, you begin sentences with an action verb instructing the reader to perform a certain task. These action verbs include:

attach	hold
begin	increase
check	install
close	join
connect	let
detach	loosen
determine	lubricate
disconnect	make
examine	make sure
expose	measure
find	monitor
finish	open
fix	place
grasp	prepare

press	select
provide	set
push	take
put	test
remove	tighten
repair	type
restore	unpack
screw	verify

Although most of the text should be written in this straightforward imperative mode, an occasional pun, joke, or other "human" interruption can break the monotony of dull instructional prose and help wake up sleepy readers. Here's an example from a manual on using PCs:

> We need just 10 of the 255 characters in IBM's extended set. Thus, we ought to be able to pack some 25.5 times more numeric information into a byte than is permitted by the ASCII coding scheme. That seems reasonable, doesn't it?

> In fact, this is exactly what is done in practice. I'm not going to put a glaze in your eyes by explaining the arcane coding schemes used—I'd have to look them up anyway!*

Although the author could possibly have explained this topic more concisely, and some readers might feel he's going off on a tangent, we like this copy; it warms up the piece, because the personality of the author comes through in the prose. While not appropriate for every technical document, an occasional shift into this more personal, more human tone can liven up an otherwise dull document for your readers.

*James Kelley, *The IBM PC Guide (Wayne, Pa.: Banbury Books)*

7. *Choose an organizational scheme appropriate to the task, the audience, and the technology.* One of the first steps in writing a manual is to decide how you will organize your material.

The organizational scheme can be presented either as an outline or in a descriptive memo explaining the way you intend to organize your document (and why).

This written proposal for your organizational scheme should be reviewed and approved by management *before* you begin writing. Reason: While technical errors and poor style are relatively easy to correct at any stage, changing the organizational scheme after a draft has been written is an expensive and time-consuming task. Therefore, an organizational structure should be selected and agreed upon before the writing begins.

The two basic organizational schemes used by instruction writers are *sequential organization* and *functional organization.*

In sequential organization, you organize your material according to the steps your reader must take to complete a task or master a subject.

Using sequential organization in a manual for a word-processing software package, for example, you might start with a few "quick and dirty" instructions so the user can immediately begin using the program to write simple memos and letters. Then you can get into more sophisticated word processing, such as footnotes, underlining, boldfacing, page formatting, page numbering, and so on.

The benefit of sequential organization is that it allows the reader to achieve quick results, without a lot of reading, and it also helps the reader master the skill faster—because it's designed to fit the way the reader thinks about the subject.

The disadvantage of sequential organization? Because the manual is designed for learning, it's an excellent tool for training novices but not such a good tool for the experienced user who is more interested in reference than instruction. Also, by being task-oriented rather than function- or feature-oriented, the sequential manual may have a lot of repetition and cross-referencing; for example, the same instruction sequence may be repeated half a dozen times or more as it applies to different tasks or operations.

Functional organization means the manual is organized around the functions or features of the system or equipment being described. A functionally organized manual for a word-processing package, for example, might have separate chapters on page formatting, text processing, text moving and manipulation, on-screen graphics, printing documents, saving documents as files, exporting files to other programs, and so on.

The functional organizational scheme has two advantages. First, because it's organized around the way the product is designed, it's easier to outline and write: You just follow the product features or functions. With a sequential manual, which is organized based on what the reader wants to know and do rather than what the product functions are, you have to spend a lot more time thinking about what exactly the reader wants to learn and the best way to teach it.

A second advantage of functional manuals is that they are better organized for reference use. Once you know how to use the system, you may prefer a functional manual because it's easy to look up and find the description of the specific function or feature you want to activate.

The disadvantage of functional manuals is that they are not ideal learning tools, because they do not present a step-by-step process for achieving specific results.

Which organizational scheme you select depends on the intended use of the manual. A manual for a training course would probably be sequential, while a reference manual for use after training might be functional. Many manufacturers produce one of each type of manual or combine them in a single document.

Strive to make your organizational scheme obvious and transparent to the reader, rather than subdued or subtle. The more apparent the organizational scheme, the easier your document is to follow.

One way to support the organizational scheme is with guideposts such as a table of contents, introduction, headings, page breaks, index, tabs, and any other devices you can use to guide the reader through your document.

The table of contents outlines all sections and subsections of the manual and gives the number of the page on which each can be found.

Heads and subheads help guide readers through the document and enable them to find information more easily. Underlining, boldfacing, indenting, and varying type size are some of the techniques you can use to define and distinguish the various levels of headings.

"Page breaks" refers to the practice of starting chapters or major sections on a new page, even if the result is some extra white space. Starting major sections on a fresh page makes them easier to find.

The index covers key terms and concepts, not every word in the manual. If a user wonders what to do when a disk is filled to capacity, he or she should be able to look in the index and find the entry *Disk, full*.

Tab dividers are especially useful in larger manuals. A tab divider for each major chapter or section allows the reader to instantly locate that chapter or section by the tab sticking out from the rest of the pages.

8. *Present instructions as a series of numbered steps.* If an operation is clearly a step-by-step procedure, you can make life easier for the reader by writing the instruction manual as a series of numbered steps:

START-UP OPERATION FOR POLYMER MIXER

1. With the injection unit fully retracted, bring the barrel and mixer to operating temperature.

2. Set the machine operation mode to manual and the boost and secondary pressure regulators to their lowest settings.

3. Quickly depress and release the injection switch. If the screw bounces back, allow more heat soaking time.

4. Once the polymer flows freely in a purging mode, increase the injection pressure as required and begin the molding operation.

With numbered steps, if the operator needs to discuss the procedure with your support staff, he or she can ask a question by referring to "step 3" instead of "the fifth line down in the second paragraph on the fourth page, where it says to release the whatchamacallit switch."

9. *Use a modular approach.* In a modular approach, sections are numbered in outline form using a hierarchical system. A heading may read:

3.4 Unplugging the spray head.

This means the material appears in section 4 of chapter 3.

The advantage of this modular scheme is that it makes it easy to delete, replace, insert, and expand sections to make corrections or update the manual as new releases or versions of the system are introduced.

For example, if the new model has an automatic water jet that washes down the spray head to prevent accumulation of solids that can plug the head, you can add a section as follows:

3.4.2. Operation of spray head self-cleaning water-jet feature.

This can be inserted as a new subsection immediately following the existing text of section 3.4. on unplugging the spray head.

According to Information Mapping, Inc., a consulting firm in Waltham, Massachusetts, that specializes in creating documentation, updating and maintaining manuals is one of the largest costs associated with documentation. Using a modular structure in which units of information can be replaced or deleted easily without having to rewrite the whole manual helps keep this cost under control and makes updating less time-consuming.

10. *Test drive your manual.* Although engineers, programmers, and system designers review manuals for accuracy, the true test of a manual's effectiveness is whether a typical user can follow it.

For this reason, we recommed that you "test drive" your manuals before you have them reproduced in quantity. This is accomplished by making photocopies of the preliminary draft, giving them to a few typical users, and getting their reactions.

Time, money, and deadlines limit the extent to which you can pretest the manual; at some point you have to create your final draft and get it published. Most manufacturers, however, err on the side of doing too little rather than too much testing of manuals.

For instance, software companies test their programs for months before release to catch every last bug, but many think nothing of putting out a manual that no one but the programmer has read and commented on. And that's a mistake. Although the programmer is the best person to catch technical errors, he or she is too familiar with the program and therefore not likely to spot explanations that aren't detailed or clear enough for the novice user.

Give your manual to a user. If he or she can follow the instructions—great. The manual is a success. If not . . . back to the word processor.

APPENDIX A

Writing in the Systems Environment

It's impossible to overestimate the influence of the computer on our society as a whole and on the writing of business documents. With the aid of software, businesspeople and technicians, who may have once balked at confronting writing problems, are now able to use computer software to help them organize, phrase, and edit their thoughts.

Throughout the book, we've referred in passing to the world of systems and to the types of documents written in that environment. Before presenting a guide to some of the software available to people who write, we'd like to sketch in a few details of the systems environment and discuss a few of the special writing issues that have attached themselves to the type of writing done by systems professionals.

Today, there are more than 1.5 million people who work in the systems departments of organizations throughout the United States—a quantum leap from 20 years ago.

Some of the chief activities within systems departments are:

• *Automation services*.This involves the use of computers to automate routine business processes such as process control, accounting, payroll, inventory, record keeping, order entry, regulatory reporting, and warehousing. The most commonly written documents in automation services include user's guides, operator's guides, training manuals, and theory-of-operations descriptions.

* *Computer services*. In this area, computers are used for such processes as management information systems, electronic mail, marketing research, and automated manufacturing. Because end users work closely with systems analysts, systems professionals need to write everything from maintenance guides to system specifications.

* *Systems services*. Professionals in this area produce documents such as proposals, systems specifications, status reports, and contracts.

Although we've mentioned these problems within the book, we'd like to briefly summarize a few of the writing issues that are especially prevalent in the world of systems:

* *Reading level*. Since user's manuals and other documents are written by technical people for those with little or no computer knowledge, there is a great premium on keeping the prose simple and using plain language to get across complex ideas. Ideally, the writer must put himself or herself into the mind-set of the users in order to write a document that will help them accomplish their goals and not overload them with details.

* *Vocabulary and acronyms*. Since the world of systems is uniquely awash in jargon, acronyms, and buzz words, the systems writer must pay special attenton to defining acronyms and using vocabulary in a clear, consistent way. For example, do you *type, depress, hit, touch,* or *press* a key? Does your reader understand that all these words indicate the same action?

* *Grammar and syntax*. Through tradition and habit, many systems professionals write in a style that blends regular sentence structure with the staccato, acronym-ridden phraseology necessary to communicate technical changes to software. While the type of shorthand used in many systems documents may be fine when writing to people with com-

puter backgrounds, it is not easily understood by the un-initiated. Therefore it helps to read your work aloud, putting yourself in the place of someone with only sketchy computer experience and asking yourself if the thought, as phrased, would make sense to such a person. For example, here is how one systems professional explained the work of his department:

Upload of following data from the UIC machines, computer initiated with no operator intervention, at product changeover.

A more easily understood way of expressing that thought would be:

At the time of the product changeover, the following data were uploaded from the UIC machines. The computer initiated this with no operator intervention.

- *Consistency.* Even though systems documents such as spec-ifications and manuals may have more than one author, there needs to be consistency in the final document. Throughout this book, we have given guidelines—on every-thing from the use of numbers to capitalization—that can help provide a uniform look to the document.

APPENDIX B

A Brief Guide to Software for Writers

If the computer has brought with it special headaches for the writer, it has also brought relief in the form of software created specifically to catch writing problems.

Although there are many word-processing packages on the market that aim at improving your writing through the use of spelling checkers, thesauruses, and format guides, we've focused here on enhancement packages directed at improving your writing. The software listings are arranged alphabetically by name and include manufacturer information and each product's features.

Be warned, however: Software does have its limitations. If we were to run great works of literature through a grammar checker, much of what makes the literature "great" would disappear. No computer software can intuit the unique needs or circumstances of a particular writing situation.

While the computer can easily catch problems such as wordiness, redundancy, sexist terms, lengthy sentences, and lengthy paragraphs, we must recognize that, until computers can be trained to think, they cannot help us very much with subtle issues of tone, subordination of ideas, the flow of ideas, and the particular way in which the sounds of words work together to dramatize an idea.

161

Here are some of the best-established software programs in the writing field:

CORPORATE VOICE
Scandinavian PC Systems
P.O. Box 3156
Baton Rouge, LA 70821-3156
IBM PC and compatibles

This program provides an ideal style model for a proposal or manual, based on an embedded database of hundreds of each kind of document. Corporate Voice moves beyond typical readability factors to match sentence cadence and rhythm, and assumes that some mechanically correct writing can be boring or unintelligible to the reader. The program can check the readability of technical documents. For example, it will keep the document at a sixth-grade reading level (or whatever reading level you designate) while compensating for the technical language.

CORRECT WRITING GUIDE
AND CORRECT WRITING
Wordstar International
P.O. Box 6113
Novato, CA 94948
DOS, Mac, Windows

Wordstar has come out with two writing guides. The Correct Writing Guide program gives the user various grammar rules. The Correct Writing program is an interactive program that allows you to activate the rules you want to apply to your writing (i.e., you can turn off rules you do not want to be checked for in your document). Correct Writing highlights problem areas and gives you suggestions on how to correct them.

GRAMMATIK V
Reference Software International
330 Townsend Street
San Francisco, CA 94107
IBM PC and compatibles, DOS, Mac, Windows

This program checks for errors in grammar, style, spelling, punctuation, and word usage; it also locates double negatives and passive language. Grammatik helps if you confuse homonyms like *its* and *it's*. The program edits the document and offers suggestions for improvement. The level of accuracy for Grammatik is high because the program analyzes words by their roots.

IDEAFISHER
Experience in Software, Inc.
2000 Hearst Avenue, Suite 202
Berkeley, CA 94709-2176
Mac Version 2.0, DOS Version 4.0

This program consists of a conceptual database of millions of characters of information, words, phrases, people, and places to help with the writing process.The IdeaFisher program asks brainstorming questions to help you come up with new ideas and associations. It is similar to a thesaurus but much more expansive, being more like a concept thesaurus—finding synonyms, opposites, and associations. IdeaFisher finds related words that may help you when creating slogans and product names.

IDEA GENERATOR PLUS
Experience in Software
2000 Hearst Avenue, Suite 202
Berkeley, CA 94709-2176
IBM PC and compatibles, DOS

This program helps you come up with ideas by asking you questions. First, the computer asks you to state the problem and your goals. Next, you are asked to think about similar situations, to find metaphors, to take other perspectives, to focus on each goal, to reverse your goals, to focus on the people involved, and to decide what to do with your ideas. Finally, you are asked to evaluate your new ideas. The program is interactive; it does not give you answers but instead asks you to come to better conclusions. Idea Generator links with Lotus Agenda and other personal information managers.

KEYNOTES WRITER'S HANDBOOK
Reference Software International
330 Townsend Street

San Francisco, CA 94107
IBM PC and compatibles, Windows, DOS

This program is designed to help with the format and organization of business correspondence. The program highlights problem areas and explains formatting and organizational problems in these areas.

PFS: BUSINESS PLAN
Spinnaker Software
201 Broadway
Cambridge, MA 02139
Windows

A step-by-step guide to writing a business plan, this package includes a word processor, a spreadsheet, and business graphics for professional-looking documents.

RIGHTWRITER
Que Software
11711 N. College Avenue, Suite 140
Carmel, IN 46032
IBM PC and compatibles, DOS, Windows, Mac

This program proofreads for errors in grammar, style, word usage, and punctuation. Rightwriter is an interactive program: Once the errors are noted and suggestions are made, you make the necessary changes. The program gives you a summary of your writing, grading you on the document's readability, use of descriptive words, and use of jargon. You can control how it analyzes your document by choosing the rules you want to apply to your writing. There are three educational levels to choose from.

THOUGHTLINE
Experience in Software, Inc.
2000 Hearst Avenue, Suite 202
Berkeley, CA 94709-2176
IBM PC and compatibles, DOS

This program links to many word-processing systems. Thoughtline is an interactive program that responds as you write. It prompts the user to supply the ingredients needed to write a successful story, such as rounded characters and live dialogue.

The program keeps in mind the construction of the document—introduction, body, and conclusion—and asks questions to help you conform to this construction.

WRITEPRO
WritePro Corp.
43 Linden Circle
Scarborough, NY 10510
IBM PC and compatibles Version 2.1, Mac Version 1.1

WritePro is an interactive program for fiction writers. The program gives general suggestions, key ingredients, that are needed in creative writing. The user interacts with the program while writing the story.

WRITER'S HELPER, Version 3.0
Conduit
University of Iowa
Oakdale Campus
Iowa City, IA 52242
Apple, IBM, Mac

This program shows different ways to restructure sentences and paragraphs. It is designed to help make your document more readable.

THE WRITER'S TOOLKIT
Systems Compatibility Corporation
401 North Wabash, Suite 600
Chicago, IL 60611
DOS, Windows

Tools to improve writing include a dictionary and thesaurus; a quotation reference; an abbreviation program; and grammar, spelling, and style checkers. The software is compatible with most word-processing systems.

Index

167

About the Authors

Robert W. Bly is director of The Center for Technical Communication (CTC), a company that specializes in improving the technical-writing skills of corporate employees and the quality of written communications within the organization. Mr. Bly holds a B.S. in engineeering, has taught technical writing at the university level, and is the author of 25 books, including *The Copywriter's Handbook*. He was formerly a technical writer for Westinghouse Electric Corporation and also spent 10 years as a self-employed technical writer.

Gary Blake is director of The Communication Workshop, a consulting firm that presents on-site seminars on business writing, technical writing, and proposal writing. Among his clients are Hughes Aircraft, American Airlines, Lever Brothers, Symbol Technologies, Chase Manhattan Bank, and Van Den Bergh Foods.

Together Robert Bly and Gary Blake have written six books, including *The Elements of Business Writing*.

Readers wishing more information can reach the authors at the following addresses: Bob Bly, CTC, 174 Holland Avenue, New Milford, NJ 07646 (tel.: [201] 385-1220). Gary Blake, The Communication Workshop, 130 Shore Road, Suite 236, Port Washington, NY 11050 (tel.: [516] 767-9590).